# A HISTORY OF A cGMP
# MEDICAL EVENT
# INVESTIGATION

# A HISTORY OF A cGMP MEDICAL EVENT INVESTIGATION

MICHAEL A. BROWN, PhD, PE

A JOHN WILEY & SONS, INC., PUBLICATION

Published by John Wiley & Sons, Inc., Hoboken, New Jersey
Published simultaneously in Canada

For general information on our other products and services or for technical support, please
contact our Customer Care Department within the United States at (800) 762-2974, outside the
United States at (317) 572-3993 or fax (317) 572-4002.

Wiley also publishes its books in a variety of electronic formats. Some content that appears in
print may not be available in electronic formats. For more information about Wiley products,
visit our web site at www.wiley.com.

*Library of Congress Cataloging-in-Publication Data:*

Brown, Michael A., 1945–
   A history of a cGMP medical event investigation / Michael A. Brown.
      p. ; cm.
   Includes bibliographical references and index.
   ISBN 978-1-118-39661-2 (pbk.)
   I. Title.
   [DNLM:   1. United States. Food and Drug Administration.   2. Adverse Drug Reaction
Reporting Systems–United States.   3. Drug Therapy–adverse effects–United
States.   4. Investigational New Drug Application–United States. QV 26.5]
   615.5'8–dc23
                    2012027158

10  9  8  7  6  5  4  3  2  1

*This work is dedicated to my wife, Mari McMenamin Brown, who made many positive suggestions in the content and preparation of this manuscript. Without her involvement this work would not have been completed.*

# CONTENTS

# PREFACE

*A History of a cGMP Medical Event Investigation* is a case study designed to introduce the U.S. Food and Drug Administration's (FDA) *Investigational New Drug* and *New Drug Application* processes, the FDA Code of Federal Regulations (CFR), Current Good Manufacturing Practice (cGMP), and the Good Automated Manufacturing Practice (GAMP) principles in the development and manufacturing of a new drug.

The case is written in a hands-on narrative format illustrating how Six Sigma principles can be applied to meet the FDA regulations in the development and market launch of a new pharmaceutical drug and medical device. Simple examples are discussed through a meeting format to illustrate both the DMAIDV and DMAIC methodologies. The program management philosophy and medical interactions are explained following implication of the drug in the death of a young woman, the transgenic mouse studies leading to drug discovery, the path to market with emphasis on shortcuts taken by a program manager with his own agenda, intracompany and government politics, quality and compliance failure, and the FDA's subsequent fatality investigation.

A level of drama is introduced to maintain the reader's interest throughout the case. The work emphasizes the essentials in bringing a new product to market through a case study of what can happen when things go wrong.

Many practices and tips for successfully managing a new product introduction are brought to the readers' attention in conversations throughout the case study. Interactions with management types who have their own agenda happen in every company and this case study will help the reader understand decisions that impact safety and effectiveness in launching a new product. These details are important and could only be learned through years of experience in a highly competitive development environment.

The case is geared toward undergraduate courses in biotechnology, engineering design, and business management and toward companies involved in biomedical and general product development. The work is self-contained, has no prerequisites, and would be an excellent choice for a company seminar that examines the federal regulations.

The case material including the FDA regulations and Six Sigma concepts was tested in an Engineering Senior Design course I taught over a three-year period at the University of Illinois at Chicago. The material is used by other professors who also teach this course. The biology and medical interactions are from course work completed at the University of Illinois College of Medicine at Chicago.

The case is fictional—this is a story to teach product development through a very probable and absorbing event that could happen. The characters though fictional exist in every goal-driven organization. One would think that there couldn't be people like these responsible for bringing new drug therapies to the healthcare market, but there have been ... and are. The science in the transgenic mouse study is current and the Agency regulations and Six Sigma principles are fundamental in many companies involved in both healthcare and general product development.

\*\*\*

I would like to acknowledge the following individuals for their valuable assistance in getting this book to press: Cover Design: Michael Rutkowski (Assistant Art Director); Editorial: Bob Esposito (Associate Publisher) and Michael Leventhal (Associate Editor); Production: Danielle Lacourciere (Associate Production Manager); and Stephanie Sakson (Production Manager at Toppan Bestset Premedia).

MICHAEL A. BROWN, PHD, PE

*Visiting Professor*
*University of Illinois at Chicago*

# PART ONE

# THE EVENT

# 1

# FRANCESCA

Francesca is a 33-year-old mother with three children: Karen, 12; Cathy, 10; and a 2-month-old son, Joseph.

Francesca is the only daughter of a well-to-do family from a Baltimore suburb. Her father, Joseph Bucco, an attorney in private practice specializing in corporation law, had a very select, elite client base. His principal client was his childhood friend Angelo Walden.

Angelo heads a family-owned waste-management business. Joseph advised Angelo on all business activities and, on a number of occasions, had successfully represented Angelo in government investigations concerning questionable practices.

Joseph passed on from pancreatic cancer ten months prior to the birth of Francesca's son. As Joseph's only child, Francesca was quite close to her father and did not take his death well, and she has been suffering from low-level bouts of depression. Her 56-year-old mother, Mary, remains in their family home and is in good health—enjoying an inheritance from Joseph's profitable legal practice and the status established from her husband's business associates and long-term friendship with Angelo. Angelo treats Francesca as if she were his own daughter and would do anything in the world to make her happy.

Francesca attended a small New England college for women in a suburb of Boston. She met John Tyler at a dance sponsored through

*A History of a cGMP Medical Event Investigation*, First Edition. Michael A. Brown.
© 2013 John Wiley & Sons, Inc. Published 2013 by John Wiley & Sons, Inc.

the women's college with a local business school in the same suburb. The two were avid skiers and enjoyed just being together—sharing an insatiable passion for licorice candy. Weather permitting they met early in the morning between their two schools for a five-mile run. Over time, the two became inseparable. John and Francesca were married shortly after graduation and returned to Baltimore, maintaining a residence in the Baltimore suburb where Francesca grew up—not far from her family home.

John was hired as a broker with the Baltimore firm Taylor and Banks and, after an unprecedented two years, was considered an expert in international investments and was promoted to the position of Senior Financial Advisor.

Through Francesca's family ties with Angelo Walden and persistent encouragement from Francesca's mother, John was introduced to principals in the Walden family business. He was given an opportunity to invest a small portion of Walden funds. The international market was exceptionally lucrative during that period and the investment more than doubled in one year. His performance opened the door for additional investment and, after six years, John was handling the total business investment portfolio. John is considered an associate in the Walden business and sees the Walden family socially playing golf with Angelo and his two sons, Anthony and Charles. John was best man at Anthony's wedding. Anthony affectionately calls John, "Johnny-boy."

Francesca is in excellent physical health. Following the death of her father and the almost immediate onset of pregnancy, she was under the care of Elisabeth Summers MD, Staff Obstetrician with the Family Health Center at Baltimore Presbyterian Hospital. She continued her regular running routine and watched her diet religiously. Her pregnancy was textbook: a full-term normal birth and a healthy infant. When Francesca was with little Joseph it brought back memories of her closeness to her father, whom she missed profoundly. Francesca was beginning to become somewhat withdrawn. Her husband noticed she was starting to avoid contact with young Joseph and had become extremely irritable. He took her to see Dr. Summers. On Dr. Summers' recommendation, Francesca began seeing Patrick Gander MD, a second-year resident in Family Practice, under the supervision of Ralph Goodman MD, Staff Internal Medicine, as well as continuing postpartum care with Dr. Summers. She was diagnosed with anxiety attacks associated with a mild form of postpartum depression. Dr. Gander, with counsel from Dr. Goodman and Dr. Summers, prescribed a new drug that showed great promise in clinical studies, with no apparent side effects, for treatment of anxiety-related postpartum depression. The drug, Oxy-Fox Inhaler, is marketed by Kinnen Laboratories.

\*\*\*

Francesca was in that state between sleep and wakefulness waiting for the 6:30 AM alarm. She didn't want to get up but she had to get the two girls ready for school, and it was a chore to even get them out of bed. Her husband was out of town on business but would be home later that evening and would help with the children.

The alarm sounded—not the low tone she preferred, but the loud, obnoxious buzzer that John needed to bring him to consciousness. She rolled over and tried, unsuccessfully, to quiet the glaring noise. Her second try met with success. She lay there for a moment collecting her thoughts for the day. She had to take her medication first, then feed little Joseph . . . he was a good baby, hadn't begun to teethe, and took his bottle well . . . and then prepare the girls' breakfast. Her breakfast would be light. Since the birth, she had begun running to knock off the extra pounds. She always loved to run and was further into her program than expected. She had reached the three-mile mark two weeks ago and today she would try for five. If she could do it, she would allow herself a treat: the licorice that John had given her before he left on his trip. She had eaten a large portion the previous night rationalizing that licorice is not fattening . . . well, maybe not.

It was a task, as expected, to get Karen and Cathy up but finally the two dressed and came downstairs. Little Joseph had his bottle while Francesca was preparing the girls' breakfast and a kind of slush of formula and cereal for the baby. The girls were running late and wolfed down their food. There wasn't time to make lunch and Francesca gave them money to eat in the school cafeterias, which the two preferred anyway. Karen was in junior high directly across the street from the grammar school Cathy attended; both schools were six blocks from their home.

John had arranged for a young Polish woman named Sophia to come over daily at 8 AM to help with the baby, do the housework, and in general watch over Francesca. John was concerned for Francesca—the bouts of depression were worrisome but the new medication was helping and though Francesca would never hurt little Joseph she was somewhat withdrawn and had ignored the baby's cries on a number of occasions. John and Francesca were close—both as friends and as lovers. Their marriage was solid.

Sophia arrived around 8 AM, actually fifteen minutes later than usual. She was in her mid-twenties, tall with light brown hair, very pretty with a good body. John had noticed; Francesca noticed that John noticed but wasn't concerned, "boys will be boys" after all. Sophia loved the children and life was much better working for Francesca than in Poland. Her English wasn't that good but she made herself understood. When excited she reverted to Polish.

Sophia apologized for her tardiness: she had missed the first bus. Francesca's expression showed annoyance with Sophia for being late; she wanted to start her run and had warmed up, stretched, and was ready, anticipating success and her licorice reward. Outward show of annoyance was out of character and Francesca associated this with her anxiety, but she thought that the medication should take care of it. She would ask Dr. Gander if the dosage could be increased on her visit this coming Friday. Francesca decided not to wait for her appointment and took a second dose—the doctors told her the drug was safe and how could it hurt.

She began the run down the street leading from her home to the forest preserve. The distance was about a half mile and there were two paths she could follow once in the preserve: the path she had run for the past two weeks gave her a total three-mile workout; the second was longer, for a total of five miles. She took the five-mile path thinking if she tired, she could stop and turn around. She also thought making the five-mile commitment would motivate her to complete the run. Her stride was long and strong and she was running a ten-minute-mile pace—she felt good at this pace, better than she had felt since the birth of little Joseph, but faster than Dr. Gardner had suggested she maintain until the full effect of the Kinnen drug was determined. She noted to herself that this pace was two miles faster than she had been running and she felt fine. Obviously Dr. Gardner was overly concerned—she felt excellent, even with the double dosage.

The five-mile path took her through a portion of the preserve that followed the river bank. She loved the path and had run it for the past ten years and only stopped in her last trimester. Francesca was thrilled to be on it again. The trees were in full bloom surrounded by patches of wildflowers and the river was alive with circulating currents. She could hear the small animals scurrying away under the foliage as she approached. Further into the run, as the endorphins kicked in, she felt even better and, without thinking, increased her pace.

Francesca finished the run in slightly less than forty-five minutes—a nine-minute-mile average. She was perspiring heavily, much more than usual, but she thought it was normal for the pace she maintained. She walked around the yard for a ten-minute cooldown, stopping to catch her breath and admire the recent work of the landscapers. She thought how beautiful her home was . . . just as nice as her family home. She thought about her father, whom she missed more than anything, and about how her father would love his namesake grandchild. She could feel her heart pounding . . . still could not catch her breath . . . still perspiring heavily. She thought it would subside.

Entering the house Sophia asked, "Fran, you okay?" Sophia could not say the word "Francesca" and Francesca abhorred being called Fran. "Fran . . . you not look good atal."

Francesca answered, "I'm alright. Need to rest a moment . . . need to catch my breath . . . somewhat dizzy . . . light-headed . . . think I'll lie down."

Francesca went up the stairs to the master bedroom. She had to cling to the railing for support. She thought about asking Sophia for help but didn't want to alarm the girl any more than she had. Moving down the hall, using the wall for support, she entered the bedroom. She loved this room. A king-sized bed was in the corner, surrounded on two sides by high bay windows. In an isolated alcove overlooking the rear garden was a desk containing her computer. A seating area with her favorite chair was in the far corner—"*God,*" she thought to herself, "*how many hours did I sit in this chair reading while carrying little Joseph?*" The bedroom was her domain . . . her comfort zone . . . the place she could get away from the problems inherent in raising two growing daughters and . . . sometimes . . . the cries of little Joseph.

Francesca lay down on the bed. She needed only a few minutes . . . once her heart settled and her breathing was under control she would take a fast shower and go help Sophia with the housework.

Her heart began to pound even harder . . . she could feel a kind of skipping followed by an even more pronounced beat. She still could not catch her breath . . . was still dizzy, even more light-headed, and was perspiring profusely. Francesca was scared. She tried to get up . . . she tried to call out to Sophia for help . . . but could do neither.

Francesca's heart did not slow down until it stopped. Francesca would never get up again.

# PART TWO

## DRUG DISCOVERY: FIVE YEARS EARLIER

# 2

# KATLIN BIOSCIENCE: TRANSGENIC MOUSE STUDY

Katlin is a small biotechnology research company with limited financial resources. The company is an offshoot of a prominent California school of medical science. Katlin was initially created as a result of the development of a nasal delivery system that could be used for a number of drug administrations normally given intravenously or intramuscularly. The delivery system proved to be successful for a few specific drugs but would not result in the profits indicated in Katlin's financial plan. Katlin has two programs in progress: to identify a replacement for estrogen therapy and to make an improvement on its initial nasal system for metering insulin. If Katlin is not successful in securing additional funding, they will have to discontinue operations completely.

William Redman PhD is President and Chief Scientist; he has a small staff of four postdoctoral students and two technicians in his organization. Dr. Redman's background is in biochemistry.

The estrogen therapy replacement investigation is near completion and has been proposed to the midsized pharmaceutical company Kinnen Laboratories for additional funding. Kinnen has shown interest but is noncommittal. The program involves the neural substrates of social behavior portrayed by the closely related monogamous California mouse and the polygynous black-footed mouse. Both mouse populations involved the development of transgenic mice with specific genes

*A History of a cGMP Medical Event Investigation*, First Edition. Michael A. Brown.
© 2013 John Wiley & Sons, Inc. Published 2013 by John Wiley & Sons, Inc.

replaced, or knocked out, that could potentially alter their behavioral patterns compared with the same populations of wild type. The wild-type (unaltered mouse) and the transgenic populations were monitored following various intravenous hormone therapies.

Development of transgenic mice could take a number of years. In the first step of the process an altered version of the specific gene to be replaced, or knocked out, is introduced into a laboratory culture of embryonic stem cells. Once the few embryonic stem cells that have their corresponding normal genes replaced by the altered gene through homologous recombination activity have been identified, the cells are then cultured to produce descendants with the altered gene expression. In the next step, the altered embryonic stem cells are injected into an early mouse embryo; the mouse produced by that embryo will contain cells that carry the altered gene. When bred with a wild-type mouse, some of the offspring will contain the altered gene in all of their cells. If two such mice are in turn bred, some of their offspring will contain two altered genes—one on each chromosome—in all of their cells. The process is laborious and can involve many generations of mouse families before the necessary numbers of transgenic mice are produced. Katlin underestimated the time involved to breed the populations and the study is behind schedule and over budget.

Past experiments with the two mouse families determined that the California mouse had substantially fewer estrogen receptors than the black-footed mouse, leading to the hypothesis that the quantity of estrogen receptors was responsible for the social behavior observed in the two species. The estrogen gene was targeted in order to provide objective evidence that modification of the gene in the two mouse populations would prove the putative hypothesis that populations with fewer estrogen receptors were monogamous and those with more estrogen receptors were polygynous. The knockout populations in the California mouse involved the elimination of the estrogen gene. The black-footed mouse population was altered so that the estrogen receptor gene is underexpressed, with the objective of altering the black-footed mouse behavior from polygynous to monogamous.

The knockout California mouse populations were split into four experiments with intravenous injections in the luteal phase of the menstrual cycle: two groups were injected, respectively, with low and high dosages of estrogen; one with oxytocin; and one with testosterone. The hormone oxytocin was included in the study because oxytocin is known to stimulate maternal nurturance and potentially improve social behavior. The transgenic black-footed mouse population was split into similar groups. The three study hormones are produced endogenously by the human endocrine system.

Dr. Redman scheduled a meeting with the two postdocs doing the investigation.

Dr. Redman opened the meeting asking Ahmed Satomi to talk them through the experimental outcomes.

Ahmed began his presentation with the knockout California mouse. "As you know the transgenic monogamous California mouse without the estrogen gene was split into four groups.

"In the group 1 population, a low dosage of estrogen injection showed that the California knockouts nurtured their young and maintained a monogamous behavior compared with the identical behavior in the wild type. These results are as expected.

"In the second group, a high dosage of estrogen injection showed completely polygynous behavior of the normally monogamous California mouse so that the offspring were abandoned and the females sought out other mates. Completely monogamous behavior was apparent in the California wild-type control. These findings support the hypothesis that low levels of estrogen are responsible for monogamous behavior.

"In the third group, injection with oxytocin maintained the monogamous behavior in both the offspring nurturance and the family social behavior of the California mouse with the knocked-out estrogen gene. Our findings identify oxytocin as a potential replacement for estrogen therapy."

Dr. Redman interrupted, "Two questions. Have you done a literature search to corroborate your findings? And what are the results of the black-footed mouse behavior."

"Yes, I have. There are a few rodent studies documented in Medline that support our outcome. But I can't find any papers that support oxytocin as a potential estrogen replacement in humans."

Ahmed continued with the experimental outcomes. "In the fourth group of estrogen knockouts, injection with testosterone in the luteal phase had little effect on the mothers but three weeks following parturition the genetic female offspring showed signs of masculinization so that they sought out female companionship, completely ignoring any males in the study group."

The other postdoc, Jim Hines, asked, "Ahmed, do these results support a basis of homosexuality?"

Ahmed laughed to himself thinking "is this a joke question?" and answered, "It was impossible to observe any sexual activity. We observed the same social behavior in the transgenic black-footed mouse population. An additional program could be developed with transgenic mice with no testosterone expression. These experiments could result in genetic male offspring with female tendencies and potential homosexual activity could be observed—two males humping would be proof of the hypothesis . . . though that would never happen."

Ahmed thought for a moment whether he should mention the last group or not. Dr. Redman had always been a stickler for following protocols exactly and the last group was not part of the protocol. He decided that the team would find the results interesting.

Ahmed went on, "The experimental protocol included only the four groups I covered. When our team monitored the first group with the low-level estrogen expression, in the wild type and the same populations in the transgenic black-footed mouse, we observed a statistically significant number in the monogamous groups that tended to abandon their offspring, withdraw from the family unit, and demonstrate aggressive actions toward their mates."

Ahmed hesitated, waiting for Dr. Redman to respond. He made eye contact and Dr. Redman told him to continue.

"An additional experiment was conducted separating these into four additional groups. Two groups of the monogamous populations of California and black-footed mice were injected with oxytocin. The other two groups were maintained as the control with no injection. The group injected with oxytocin returned to nurturing their young but maintained the aggressive tendencies toward their male partners. There was no change in the control group—these two groups continued to ignore their offspring and demonstrate aggressive tendencies toward their mates."

Dr. Redman was quiet for some time. Ahmed waited for his response. Dr. Redman began, "I'm extremely interested in your data. We need to get together later and discuss if we can take the study with the two groups one step further in adding a very low dosage of an antianxiety drug to determine if the aggressive behavior of the two populations can be controlled."

Ahmed smiled, "I thought you would find this as interesting as we did." He continued with the remainder of his talk. "For the sake of time I'll only summarize the studies with the low estrogen gene expression transgenic black-footed mouse.

"We had four populations. Each responded with similar behavior modifications as the California mouse populations. The low-level estrogen population demonstrated a switch from polygynous behavior to monogamous; the population injected with an increased dosage of estrogen maintained polygynous behavior; injection with oxytocin maintained the monogamous behavior in both offspring nurturance and family social behavior, supporting the results of the California mouse population; the testosterone injections resulted in offspring masculinization, and a small population of the monogamous groups demonstrated the identical antisocial behavior that I just covered."

Dr. Redman was pleased with the work of the two postdocs. He complimented them and said they needed to get together later.

From Ahmed's investigations it was apparent that Katlin had evidence to support the role of estrogen receptors in the social behavior of mice. The data supported the hypothesis that oxytocin could be a potential replacement for estrogen therapy—though Dr. Redman had reservations on the effectiveness in humans. Oxytocin has been well characterized in humans, with no indication of estrogen-like expression. This study would not bring in the funds Katlin needed.

Dr. Redman had little comment on the data concerning masculinization tendencies and didn't want to pursue a course of study. He told his staff the masculinization research was not part of the current experimental protocol but could be part of a future study and he'd follow up with colleagues at the university, but the testosterone data must not be included in the experimental Redbook. Research in homosexuality, if someone would actually take it seriously, could get the right-wing Christian organizations on his case—though the publicity could bring much-needed attention to their research.

Dr. Redman was extremely interested in the mouse data with behavioral modifications in the care of their young as well as the female withdrawal and aggressiveness toward their mates. There was a market for a drug that could help the many mothers who demonstrate postpartum issues. But the team needed to demonstrate an effective modification that would control the aggressiveness tendencies. Dr. Redman met with Ahmed later that day, telling him they had to think about how additional studies could be done and added to the experimental protocol.

The proposed studies raised serious issues: An extended study with an antidepressant or antianxiety drug should require a new transgenic mouse population to be bred, with a completely new preapproved investigation plan. The company didn't have funds to do so and wouldn't be in existence by the time a program could be funded. The potential of discovery, even if serendipitously, of a therapy that relieves postpartum issues could be financially immense. Also on Dr. Redman's mind was how such a cocktail could be administered. Antidepressant drugs are lipophilic and can be taken orally, but oxytocin is a peptide and wouldn't be absorbed by the digestive tract. He was aware that oxytocin, when used for "milk letdown" in nursing mothers, is given as a nasal spray . . . even though a nasal spray device is the company's expertise, he was not so sure about nasal administration of an antidepressant.

His first choice was a benzodiazepine. Benzodiazepine is an anxiolytic drug prescribed for the treatment of anxiety and is safe and effective in humans. Dr. Redman didn't know of any mouse studies along this line but, if the aggressiveness could be controlled, it could lead to a marketable oxytocin–benzodiazepine cocktail to use as a postpartum antidepressant and potentially secure the funding Katlin needed.

Dr. Redman consulted with his university colleagues to pick the actual benzodiazepine he should use in the studies. After a week's review the university told him that, for cases of postpartum depression, a benzodiazepine isn't normally prescribed. A class of antidepressants known as selective serotonin reuptake inhibitors (SSRIs) is the first-choice medication for treatment. Most SSRIs are considered relatively safe for use while breast-feeding because in general they pass into the breast milk at very low levels. Side effects are minimal but may include various degrees of sleep disturbances and sexual dysfunction, and an overdose can cause seizures.

His colleagues advised him that the university's Pharmacology Discovery Department had recently received approval on a new tricyclic compound called foxepin. Tricyclics are rarely used because overdose can cause toxicity and reflex tachycardia—though only an insignificant animal population of the foxepin preclinical participants given the drug actually demonstrated tachycardia, and there were more cases of tachycardia in the placebo group. More important for treatment of postpartum depression the preclinical studies demonstrated that, in the case of mixed anxiety–depressant episodes common in many cases of postpartum depression, foxepin was more effective than SSRI treatment—without the toxicity and adverse side effects demonstrated with other tricyclics. As a class of tricyclics, foxepin could also be safe when breast-feeding. Foxepin is a lipophilic compound and is normally taken orally but, if incorporated into a single molecule with the peptide oxytocin, could be administered through the nasal passage.

Dr. Redman's colleagues stressed that the university would require his rodent study to use the newly developed foxepin. The university had applied for the patent rights and it would be the licensing strategy. All other forms of treatment were off patent and in the public domain. The university was excited about the research and expected a licensing agreement resulting in a substantial profit.

Dr. Redman decided to go ahead with the current experiment using the existing transgenic populations—treating the withdrawn and aggressive mice with a very low-dosage foxepin injection. He didn't have the time to follow a new extended study and had to do the modified experiments with the existing mouse populations or his company would dissolve. He figured that foxepin administered directly into the mouse bloodstream would mimic the absorption by the nasal passage in humans, and he had no knowledge how to do a nasal aspiration on a mouse.

The preclinical studies conducted by the university showed that oral dosages of foxepin were safe and effective in two animal models and there was no indication that there would be contradicting results during Phase I human toxicity studies. As a matter of fact tricyclics are well characterized in human studies and should not require additional toxicity studies.

The results proved to be successful; both oxytocin mouse groups following the foxepin injection not only continued to nurture their young but rejoined the family unit seeking out their mates. The experimental documentation would be updated by addendum with the new data.

Kinnen Laboratories expressed interest in the estrogen replacement therapy investigation but Dr. Redman knew that oxytocin would not result in a viable replacement therapy. He was confident that the oxytocin–foxepin nasal spray would really get their attention.

Dr. Redman met with his colleagues the next morning to review the animal testing documentation that the due-diligence team from Kinnen would require before licensing the new drug.

Dr. Redman met with Ahmed and began the meeting with a brief statement, "As I am sure you know, if Kinnen licenses the compound, they will require certain documentation that must be included in the Food and Drug Administration's Investigational New Drug Application. I know we had to take some shortcuts in the experiments but still," he paused, "where do we stand?"

Ahmed hesitated, apparently thinking of the best way to answer, "Dr. Redman, as you know, the animal studies are designed to test ranging dosages of the compounds in order to obtain preliminary efficacy and pharmacokinetic information. We agreed that due to time limitations a new protocol could not be generated and a new mouse population would be financially out of the question. Now, as you are aware we made shortcuts in testing."

Dr. Redman said, "Please be specific; I need to know exactly what we did not do."

"For one, we used the same mouse population that indicated a positive response to oxytocin in the behavioral studies. As the experiment was directed to study a new compound, we should have developed a new protocol with a new population, with the oxytocin and foxepin injected at the same time. Second, when we administered the foxepin antidepressant intravenously into the mouse bloodstream, the oxytocin concentration would have degraded over the time prior to the injection. So we have no definitive evidence what concentrations of oxytocin were in the mouse populations relative to the dosage of foxepin. Third, and most important, we did only one injection with one dosage. We should have included dose-ranging in our study to determine the safe limits of the drug. We do not have data that will satisfy this requirement; who knows, maybe a milligram more could have been fatal. And the Agency requires toxicity testing studies on at least two animal species: obviously not part of our studies. We know from the university that the preclinical animal studies resulted in a small percentage of dogs that developed tachycardia, but this data cannot be used in our mouse studies to satisfy any of the toxicity requirements."

Ahmed hesitated again; in his mind he was not sure where this conversation was going, "Dr. Redman, just to be clear, the Agency has very specific and enforceable regulations for 'Good Laboratory Practice' as described in the *Code of Federal Regulations*, 21 CFR 58 on the Agency website—www.fda.gov. At a minimum we need to provide job descriptions and training documentation for all our people engaged in the studies and to provide objective evidence that all of the reagents used on the mouse populations were tested for identity, strength, purity, stability, and uniformity; also, we did not maintain the appropriate quality control documentation required. Now, in my opinion, the administration of the two drugs independently does not meet Agency requirements. I believe that a carrier will have to be identified that will bind the two molecules."

Dr. Redman interrupted, "We did maintain documentation signed off by the technician and me. This should meet the Agency's requirements."

"No. You are not recognized as the Quality Control Unit in Agency terms. You are the Study Director. And we do not have a Master Schedule describing all the studies we conducted. Not to even mention that the main experiments were conducted after the fact without preapproved protocols.

"In addition, we should follow an accepted quality method applying Six Sigma fundamentals in developing a clear definition of the program goals; a specific plan of the genetic design requirements, or 'Critical to Quality' goals noted as CTQs, to meet the program goals, noted as the 'Voice of the Customer'; preapproved validation plans for the evaluation of the rodent populations; testing protocols, and the resulting data organized in such a manner that it is obvious that the results meet the CTQs."

Dr. Redman responded, "Following the Six Sigma methodology to document our experiments is not an Agency requirement and I do not want to discuss this again. I know what I am asking now may be considered a violation of ethics, but can you 'dry-lab' the experiments?"

Ahmed said, "You mean change the experimental protocols to include the additional testing and enter the results post preapproval? And prepare the quality records now?"

"Yes, that is exactly what I mean."

# 3

# OXY-FOX INHALER

## 3.1  KINNEN LABORATORIES

Kinnen is a diversified healthcare company with three divisions: Pharmaceuticals, Medical Devices, and Therapeutics. Kinnen is a medium-sized player in the healthcare market with a few highly profitable pharmaceutical drugs in areas including cytotoxic and hormonal drugs for chemotherapy, immunology, and chronic pain relief. The Medical Devices business is relatively new and is growing mainly through acquisitions. Therapeutics is well established so that the business is more or less in a maintenance mode. Immunology is a subset of the Pharma Division and is a cash cow with franchise drugs marketed worldwide. Kinnen is focusing its next line of drugs in the neuroscience and paincare pharmaceutical market.

The initial push was in erectile dysfunction (ED) and in a newly discovered toxin that could provide pain relief. The toxin is taken from the skin of a specific Amazon toad used by tribes to incapacitate their prey.

The ED drug was identified by hospital emergency room doctors treating cases of ingested toxins. The emergency room medication is a cocktail of a diuretic and a gastrointestinal irritant acting on the peripheral afferent neural pathway inducing vomiting. Following

*A History of a cGMP Medical Event Investigation*, First Edition. Michael A. Brown.
© 2013 John Wiley & Sons, Inc. Published 2013 by John Wiley & Sons, Inc.

administration of the drug, the patient immediately vomits—and many times quite violently. The doctors recorded that in a large population of male patients a very distinct erection was noticeable. Kinnen's research team designed a derivative of the cocktail that proved to be psychogenically effective through receptor binding of the putative neural pathway leading to male sexual arousal. The drug was not approved domestically for the treatment of ED due to a rather high clinical identification of the annoying side effect of continuous and violent vomiting. The drug was approved internationally by the European Regulatory Agency, obviously without regard to women's preferences in those countries.

The toad toxin, identified as DX-32 in Kinnen studies, was potentially a replacement for the multibillion-dollar morphine market. The toxin has the same analgesic effects as morphine without the potentially severe side effects and dependence. DX-32 failed in animal studies as the toxin potency had to be held to a 100-nanogram ($10^{-7}$ g) dosage to avoid sudden death—a difficult manufacturing restriction. With these two failures, Kinnen is aggressively seeking new drugs to support growth in the Neuroscience and Pain Care business.

## 3.2   KINNEN LABORATORIES: OXY-FOX TRANSFER

Tom Watson, MD PhD, is Vice President (VP) of Pharma Research and Development (R&D) reporting to the Pharma President, R.L. Siegal. The Research and Development organization comprises programs focused on Immunology; Oncology; Neuroscience and Pain Care; and Licensing and New Drug Development. Each program is headed by a Principal Scientist.

Dr. Watson has been with Kinnen for over 20 years, starting in the Therapeutics Division and transferring to Pharma as the Principal Scientist in Licensing and New Drug Development. His success as a scientist, as well as his terrific people skills, paved his way to heading the research area. Dr. Watson has been VP for over a year. He plays company politics only to the extent of protecting his organization.

With emphasis on the growth of the Neuroscience and Pain Care product line, the R&D organization is investigating drug precursors internally as well as supporting research in a number of biotechnology companies and universities. The most promising drug is Oxy-Fox, an oxytocin–foxepin antidepressant for postpartum depression developed by Katlin BioScience. Through a licensing agreement with Katlin, Kinnen gained drug development and marketing rights for the antidepressant. Oxy-Fox has the potential to establish a billion-dollar postpartum antidepressant market.

## 3.3   DUE-DILIGENCE TEAM AND KATLIN DATA ACCEPTANCE

Dr. Watson negotiated the terms of the license personally with the assistance of the Finance VP but left the due-diligence technical review to a young biologist in his organization, Jeffrey Daniels PhD. Jeff shares the responsibility to accept the report with individuals from Program Management, Product Development, and Materials Management. The due-diligence final report indicated that the transgenic rodent study was effective and the data statistically viable. Kinnen feels that the drug is appropriate for human testing and has applied to the Food and Drug Administration (FDA; or simply "the Agency" for short) for an Investigational New Drug (IND) application. Jeff sees the successful transfer of this drug to Program Management as his ticket to further responsibility in the R&D organization. Jeff is somewhat naive and was heavily influenced by the Program Manager assigned to the team in the acceptance of the rodent study data.

Keith Carlisle represented Program Management and was assigned to manage the Katlin BioScience due-diligence team. Keith reports directly to Larry Fletcher, Pharma VP of Operations, and is his "go-to guy" for high-priority programs. The postpartum antidepression drug is a major goal for the Pharma Division—probably the most visible program in the corporation—and Keith is confident that he'll manage the clinical studies and market launch following the acceptance of the due-diligence report. Keith will be successful and will overcome any obstacle that could delay the program. The launch of this product will be the turning point in his career and will guarantee fast-track promotion. Keith plays the intra-Pharma politics and has done quite well. Following animal studies, the normal course for human trials and clinical evaluations is a minimum of five years, usually seven to ten. Keith's personal goal is to cut it to no more than two years. He plans to use the rodent study data to reduce the approval time and petition the Agency for "fast-track" approval based on patient need. Keith has recently completed an MBA from a local college and considers it the first step to his next career move—to Director of Program Management. Keith will not allow the market launch of the new postpartum drug to follow the usual course of seven to ten years. His career plan has him in the Director role within two years.

Gordon Taylor, Associate Principal Scientist, was assigned as the Product Development representative on the Katlin BioScience due-diligence team. Gordon reports to Edward Chase, Director of Product Development. Chase reports directly to Larry Fletcher, Pharma Operations VP. Gordon has been in the Product Development organization for a number of years and, though educated in organic chemistry, has little experience in animal studies or Agency submittal requirements. He has never been outspoken and avoids conflict whenever possible. Gordon is a team player and is recognized as a scientific contributor who can be counted on to successfully complete his team assignment.

Both Gordon and Jeff identified a number of discrepancies in the rodent study documentation that could prove to be troublesome. The discrepancies were discussed at their team meeting but the Program Manager convinced them that the statistical data was more than viable. Though Jeff was the scientific team leader, he sensed that any objection would be a no-win situation and went along with the direction of the Program Manager. Gordon kept quiet. Gordon could be a significant influence if he shared his expert opinions and adopted more of a leadership role.

Dave Stall, Group Manager Sourced Materials, was assigned as the Materials Management representative on the Katlin BioScience due-diligence team. Dave reports to Jennifer Feddler, Pharma Director of Materials Management. Jennifer reports to Larry Fletcher. Dave's responsibilities were strictly focused on identifying a qualified supply line for the raw materials required for market launch and sustained postmarket production. Though a member of the team, he was not allowed to participate in the technical evaluation of the study data. However, Dave is a conscientious manager and reviewed the rodent study documentation; he identified experimental discrepancies from a program organization point of view that required explanation. Dave was told by the Program Manager, Keith Carlisle, that he did not have the background to fully understand the experimental outcomes: The review and acceptance of the Katlin data was not his responsibility.

# PART THREE

## KINNEN OXY-FOX INHALER MARKET LAUNCH PROGRAM

# 4

# AGENCY IND AND NDA REQUIREMENTS, SIX SIGMA CHARTER, AND DEVICE MASTER RECORD

## 4.1 LAUNCH TEAM MEETING NUMBER 1

### In Attendance

Keith Carlisle, Program Manager—responsible for the Oxy-Fox Inhaler market launch

Jeffrey Daniels, Research and Development (R&D)—responsible for the development of the Oxy-Fox compound

Susan Jaffey, Quality and Compliance—assigned to be Quality Lead on the Oxy-Fox Inhaler launch team; reports to Marcia Hines, Pharma Director of Quality and Compliace

Dan Garvey, Group Manager Support Engineering—responsible for all manufacturing equipment design and fabrication, biotechnology, polymer engineering and plastic production, and product manufacturability

Gordon Taylor, Associate Scientist Product Development—lead scientist responsible for the plastic component design for the Oxy-Fox Inhaler

Dave Stall, Group Manager Sourced Materials—responsible for all materials required for new product development and market launch

*A History of a cGMP Medical Event Investigation*, First Edition. Michael A. Brown.
© 2013 John Wiley & Sons, Inc. Published 2013 by John Wiley & Sons, Inc.

Jim Gonzales MD, Medical Affairs—responsible for managing all Oxy-Fox Inhaler submissions to the Food and Drug Administration (FDA; or simply "the Agency" for short) and clinical studies

Following the acceptance of the due-diligence report by division management, Keith Carlisle, as he expected, was chosen as the Program Manager for the product launch. Each director assigned representatives from their respective organizations to support this effort. The team consisted of Jeff Daniels from R&D, Gordon Taylor from Product Development, Dan Garvey from Engineering, Dave Stall from Materials Management, and Susan Jaffey from Quality. Initially Dave and Dan would represent the plant function and Dan would work with Jeff Daniels in the development of the benchtop Oxy-Fox process development and the transfer to the biotechnology group for manufacturing. In addition, Keith Carlisle would work closely with Jeff Gonzales in managing the Agency submissions. Dr. Gonzales would attend all team meetings.

In the first team meeting Keith presented an agenda and explained that the focus of the meeting would be the review of individual schedules. First on the agenda were introductions of the team members, followed by Dr. Gonzales who would present a brief outline of the Agency process with a clinical timeline; then Sue Jaffey would review the quality initiative for the product launch, and Dan Garvey and Gordon would review the major paths their groups will follow to develop the Oxy-Fox process and nasal inhaler design.

Following introductions, Dr. Gonzales told the team he would present an overview of Agency policy on the submittals process but could not present a timeline yet. Keith Carlisle immediately interrupted, saying that he expected a schedule from Medical Affairs, not just an explanation of Agency policy. Dr. Gonzales knew of Keith's reputation and decided to ignore him for the time being and continue.

Dr. Gonzales began, "I've worked with each of you previously and I'm sure that you're aware of the Agency's requirements . . . please bear with me.

"The plan is for clinical studies to be done at three sites in two different states. Before the Oxy-Fox Inhaler can be transported or distributed across state lines we have to have an exemption in the form of an 'Investigational New Drug,' or IND in Agency abbreviations. The IND application will contain information in three areas: animal pharmacology and toxicology studies; manufacturing information; and clinical protocols and investigator information. The IND is being prepared for submittal and will require some information from this team.

"The animal pharmacology and toxicology studies contain preclinical data to permit an assessment as to whether the product is reasonably safe for initial testing in humans. Also included are any previous experiences with the drug in humans; we'll use the data from Katlin as well as a number of published human studies. The animal studies are designed to test ranging dosages of the compounds in order to obtain preliminary efficacy and pharmacokinetic information. When I spoke with Jeffrey Daniels, the lead scientist on the due-diligence report, he told me there were some discrepancies in the animal data. He didn't go into details but this is a concern."

Dr. Gonzales continued, "The manufacturing information addresses information pertaining to the composition, manufacturer, stability, and controls used for manufacturing the Oxy-Fox compound and the delivery system. This information is assessed to ensure that we can adequately produce and supply consistent batches of the drug. Preliminary information from both Gordon and Dan is required as part of this submission.

"The clinical protocols and investigator information assess whether the initial-phase trials will expose subjects to unnecessary risks. Also, information on the qualifications of the clinical investigators who'll oversee the administration of the experimental compound are required to assess whether they are qualified to fulfill their clinical trial duties. I'll manage this part of the submission with our clinical investigators. Once the IND is submitted, we must wait 30 calendar days before initiating any clinical trials. During this time, the Agency has an opportunity to review the IND for safety to assure that research subjects will not be subjected to unreasonable risk. Both oxytocin and tricyclics have been characterized in previous human studies, so this may not be an issue.

"To be sold legitimately in the United States, new drugs must pass through a rigorous system of approval specified in the Food, Drug, and Cosmetic Act and supervised by the Agency. No new drug for human use can be marketed in this country unless the Agency has approved a 'New Drug Application,' known in Agency terms as an NDA. The documentation required in the NDA describes the drug's history, including the outcomes of the clinical tests, the ingredients of the drug, the results of the animal studies, how the drug behaves in the body, and how it is manufactured, processed, and packaged—including full validations for the equipment and facilities in which the product is produced; a sampling and testing plan of in-process materials and final products; and appropriate written procedures, designed to prevent microbiological contamination of drug products purporting to be sterile."

Dr. Gonzales paused, looking around the room. "Are there any questions at this time?"

With no questions, Dr. Gonzales continued, "Trials in the first phase, noted as 'Phase 0,' include the administration of single subtherapeutic doses of the Oxy-Fox investigational drug to a small number of subjects, no more than 15, to gather preliminary data on the agent's pharmacokinetic and pharmacodynamic properties and mechanism of action. The results from this phase are usually included in the IND. The Phase 0 information will be taken from on-market similar drugs and is reported as complete—though, as I mentioned, I've reservations with the extent of these studies as pointed out by Jeff and if on-market studies are even appropriate. These studies are required to discover the degree of any interference and the extent of any toxic effects. Once I'm confident that there are no interferences or toxic effects my group will prepare a detailed plan for the proposed human testing."

At this point Keith Carlisle interrupted, "Jim, the due-diligence team approved the study and was obviously convinced that the studies where clinically viable. This is a dead horse and we should proceed with full clinical trials. As Program Manager I would like to review your plan. Do you have a plan?"

"Keith, this is not the time to debate the merits of the reported animal study. I've reservations, as I pointed out. And I don't have a plan yet. You've done this before and should know that we cannot even begin clinical studies until our drug is produced and the delivery system is available. Let me finish explaining the clinical process.

"Drug testing in humans consists of three consecutive phases. 'Phase I' trials are the first stage of testing in human subjects. Normally a small group of twenty to eighty healthy volunteers are selected. This phase includes trials designed to assess the safety, tolerability, and pharmacodynamics of the Oxy-Fox therapy. Phase I testing is most often done on young, healthy adults. In the case of the Oxy-Fox Inhaler this testing will be done in an Agency-approved obstetric program with a relatively small number of subjects diagnosed with postpartum depression symptoms. The purpose is to learn more about the biochemistry of the drug: how it acts on the body and how the body reacts to it. Phase I trials normally include dose-ranging studies so that doses for clinical use can be refined. The tested range of doses will usually be a small fraction of the dose that caused harm in animal testing.

"And to repeat myself, these studies can't even begin until we have a supply of nasal inhalers produced on a validated process that will be used for final product manufacturing.

"Once the initial safety of the Oxy-Fox drug has been confirmed in Phase I trials, Phase II trials are performed on larger groups and are designed to assess the activity of the drug, as well as to continue Phase I safety assessments. The development process for a new drug may fail during Phase II trials due to the discovery of poor activity or toxic effects. Some trials combine Phase I and Phase II into a single trial, monitoring both efficacy and toxicity.

"During Phase II, small controlled clinical studies are designed to test the drug's effectiveness and relative safety. These will be done on closely monitored patients who suffer from various degrees of postpartum depression. Some volunteers for Phase II testing who have severely complicated conditions may be excluded. A 'control' group of patients with comparable states of depression or anxiety will be used in double-blind, controlled experiments. These experiments are conducted by medical investigators thoroughly familiar with this condition and this type of research. In a double-blind experiment, the patient, the health professional, and other personnel will not know whether the patient is receiving the Oxy-Fox drug, another active drug, or no medicine at all . . . a placebo. This helps eliminate bias and assures the accuracy of results. The findings of these tests are statistically analyzed to determine whether they are 'significant' or due to chance alone.

"Phase III consists of larger studies. This testing will be performed after effectiveness of our nasal spray has been established in the preceding two phases and is intended to gather additional evidence of effectiveness for specific use. These studies also help discover adverse drug reactions that may occur. Phase III studies may involve a few hundred to several thousand patients who suffer from this condition."

Keith Carlisle could not contain himself any longer and stopped Dr. Gonzales. "Dr. Gonzales, please do not misunderstand me; this program is a top priority for the company and I cannot accept the time frame that you are suggesting . . . I mean, several thousand patients is ridiculous. My initial point was I wanted to review a plan to include overall timing of the studies. Your presentation is educational but we need this meeting to review schedules from the Quality organization, Product Development, and Engineering as well, not a detailed talk on how drugs are accepted by the Agency. You pointed out that the inhaler must be designed and fabricated as well as the Oxy-Fox cocktail process developed and scaled up for production. We need to review these plans as soon as possible and in my opinion we're wasting time"

Dr. Gonzales responded, "Keith, I mentioned that the clinical studies can't begin until the inhaler is available. I understand the team needs to focus on the design and development of the cocktail and inhaler. The team also needs to understand the importance of the clinical studies and their part of the NDA submission. The clinical studies are on the critical path for market launch only because completion relies on availability of the inhaler. Let me continue without interruption.

"The inhalers don't have to be made in final manufacturing. The Oxy-Fox Inhalers used in clinical studies can come from prototype processes as long as the final manufacturing processes are shown to be equivalent to the prototype processes. The clinical studies can be done on a parallel path with production scale-up. But, to repeat myself, the prototype production process is required to be validated. I'm almost done.

"In addition, the normal process involves the inclusion of patients with additional diseases or those receiving other therapy indicated in the product labeling. They would be expected to be representative of certain segments of the population who would receive the drug following Agency approval. In our case, the inhaler is specific to postpartum patients with symptoms of child abandonment and irritability as well as general depression, and this may not be necessary.

"But remember a few years ago, a pregnancy test marketed by one of our competitors was used by a physician on a woman in her early twenties who was not pregnant. The patient's test results indicated a high HCG hormone level, or human chorionic gonadotropin, and based on this and the fact that the woman was not pregnant the physician diagnosed the patient to have cervical cancer. The test is not indicated in the product insert as a cancer screen, but the insert did not specifically say that the test could not be used to test for cervical cancer. Though males with testicular cancer do express high levels of HCG in late stage and there is an approved test for this condition, women do not express a peak in HCG until early in the second month of pregnancy . . . or if they have testicular cancer . . . which normally is not the case. Without any confirmation testing for cervical cancer and no additional testing to determine if the initial assay was a false positive or why the patient expressed the high HCG level, the young woman was subjected to a complete hysterectomy she did not need. The Agency held the company responsible because their insert did not specifically say that the test should not be used as a cancer screen. Since then our entire industry has been gun shy.

Dr. Gonzales paused for a moment looking around the room to register if the team understood, then continued, "When the clinical studies are complete and the obstetrical–gynecological—OB-Gyno— programs believe the investigational studies on the nasal inhaler have shown it to be safe and effective in treating the condition, the NDA is submitted to the Agency. This application is accompanied by all the documentation from our research, including complete records of all the animal and human testing. The NDA and its documentation must then be reviewed by the Agency's physicians, pharmacologists, chemists, statisticians, and other professionals experienced in evaluating new drugs. Proposed labeling information for the physician and pharmacist is also screened for accuracy, completeness, and conformity to Agency-approved wording.

"Regulations call for the Agency to review an NDA within 180 days. This period may be extended if additional data is required and, in some cases, may take several years. When all research phases are considered, the actual time it takes from idea to marketplace may be eight to ten years or even longer. However, for drugs representing major therapeutic advances, the Agency may 'fast-track' the approval process to try to get those drugs to patients who need them as soon as possible. You may recall that we have received fast-track approval for two of our Immunology drugs. In our case, since the Katlin studies demonstrated that there were no toxicity effects you can figure that the approval process could be complete as early as three years and could take possibly up to six. I do not see the Agency granting fast-track approval.

"And after the inhaler is marketed, we must inform the Agency of any unexpected side effects or toxicity that comes to our attention—this is referred to as Phase IV in Agency terms. Consumers and healthcare professionals have an important role in helping to identify any previously unreported effects. If new evidence indicates that the drug may present an imminent hazard, the Agency can withdraw approval immediately and require us to remove the inhaler from the market or require new information to be added to the drug's labeling. In either case the product is off the market."

Again, Dr. Gonzales paused to make sure the team understood his last point; he then continued, "It's in our best interest from a company perspective to complete this process according to the policy I've described. Any shortcuts, even if Agency approved, could lead to field problems and the inhaler could be recalled. This is not something we want to happen.

"In addition to these phases of testing for the Oxy-Fox Inhaler, we may have to follow the 'premarket approval' process for the inhaler, known as a PMA for medical devices. Not absolutely sure on this and have requested an interpretation from our consultants."

Dan Garvey interrupted, "Isn't the Katlin nasal inhaler on the market and doesn't it have the same characteristics as the device we're designing? So if we have to follow the medical device submittals why do we have to follow the PMA regs and not just a 510(k)* submittal for equivalency of the already approved inhaler?"

Dr. Gonzales answered, "Dan, not sure . . . that's why I'm seeking an interpretation. A major requirement of the nasal spray inhaler is the metering feature at the tolerance required for the administration of the foxepin, which I don't believe is approved for the Katlin device or any on-market device. In either case the documentation required for product development and engineering submittals will be the same."

Keith asked, "Dr. Gonzales, are you done?"

"Yes. I've covered an outline of my group's responsibility. Keith, I have to go . . . take a minute and walk me down the hall."

\*\*\*

As the two walked to the elevator, Keith showed visible signs of annoyance but kept his mouth shut. Dr. Gonzales said quietly that he would set up a meeting as soon as possible between them and his boss to discuss how the program could be fast-tracked within Kinnen, potentially using literature studies on both oxytocin and tricyclics to satisfy the clinical toxicity study requirements. He assured Keith that his goals were the same as Keith's, though from a medical responsibility perspective it must be documented that the inhaler meets the Agency's safety requirements. Keith agreed and went back to continue the first team meeting, thinking that at this pace his career might just follow the normal eight- to ten-year course after all and his promotion to Director would be just as far off—if ever at all.

\*\*\*

Back at the meeting, Keith referred to the meeting agenda showing Susan was next, to review quality procedures, and then Gordon would address the design of the inhaler plastic components, and Dan the manufacturing process. Keith went on to say that they didn't need any more discussion on Agency requirements. Sue took the floor and began.

"Before I discuss the quality function for this program I want the meeting minutes to note that the due-diligence team did not have a Quality representative." Keith immediately interrupted, "Sue, the Katlin team was selected by Pharma upper management; if you have an issue, bring it up with your manager."

---

*FDA, Submission for Medical Devices (listed in Bibliography).

"I will do that. Now, before the NDA can be approved for the inhaler, we must demonstrate to the Agency that all requirements relating to production are met. These include the Agency's 'current good manufacturing practice' regulations, abbreviated cGMP, and any applicable standards relating to strength, quality, purity, packaging, and labeling that are established by the United States Pharmacopeia. I am sure you all recognize this organization name as USP. And . . ."

Keith stopped Sue in midsentence, "Sue . . . this is more Agency stuff. We don't have time for this discussion."

"You let Gonzales finish; now let me. I'm leading to a point."

Sue disregarded Keith's objection, continuing, "Our group will do testing required by Agency cGMP regulations. The Agency does not routinely test all products, except in cases where there's a suspicion that something may be wrong and this is the last thing we want as an organization. In addition to governmental requirements, drug products must meet public standards developed by USP as I mentioned. In order to market the inhaler in the United States we must meet USP established standards unless we specifically choose not to meet the standards. In this case, the inhaler label must state that it is 'not USP approved' and how it differs from USP standards. This will not be the case; we will meet USP standards.

"I've also been instructed by my boss that this program will follow Six Sigma policy for the design and production of the inhaler and Oxy-Fox process. I've spoken with both Gordon and Dan and they agree."

Keith nodded in agreement and said, "Sue, both Dan and I are Six Sigma Black Belts so this won't be an issue. Between the two of us and with Gordon's input we can get this done. What part of this documentation will Quality be responsible for?"

Sue answered, "Quality will be responsible for only review and approval; we'll not write the documents for you, but I'll be available to discuss any experimental deviations or issues and how to handle additions to the document package to account for these.

"Let me continue with the documents we want to see in this package.

"Whether we need to follow the Agency's medical device requirements or not, the documentation for the inhaler design and the process development will follow the regulations under design controls. My group will be responsible for organizing the 'Design and Development Plan' and maintain the 'Device Master Record,' identified as the DMR in Agency regulations.

"First a Project Charter is required by our Six Sigma program, in the form of a proposal used by both the team and management to judge our progress—and of critical importance is that approval of this document assures management support. I have my laptop set up and am putting the form we need to complete on the overhead projector, and I'll pass copies out for your review. As an example, I've filled in some of the information. We need to set up meetings off-line to gather the information." (See Fig. 4.1.)

Sue continued her presentation, addressing each part of the charter.

"Looking at the overhead, the charter identifies the customer, in our case this is Marketing; the project team members are listed with the responsibilities of each member and approval by their organizational Director for participation; this is followed by the project objectives, which must be defined in measurable terms; the definition of when the program is complete includes the dates when the inhaler must be available to begin clinical studies and the scheduled date to launch, as well as final sign-off on all documentation at a minimum; key deliverables are listed, such as when the preliminary Oxy-Fox process is finalized and qualified, and when the prototype assembly line is available and qualified; key assumptions and conditions are listed; I do not expect to see any assumptions—assumptions must be resolved; an executive-level timeline and major milestones must be identified by the project engineers and design team; all preliminary cost estimates requiring updated costs when finalized; a list of design inputs, which are the customer requirements, and the design outputs, or design specifications, to meet the customer inputs; summaries of both the verification and validation plan for both the inhaler components and the manufacturing process; a list of major risks that can affect the project and a brief plan to address these risks—an example of a major risk is if the automated assembly equipment does not work, in which case Engineering must identify a plan to keep the program on track while the equipment is being redesigned; identification of all documentation with a brief description and who is responsible for completion of the documentation, and when the documentation must be complete, and by complete I mean reviewed and approved by the Quality representative, which is me, and any risks in completing the documentation."

Dan Garvey asked, "Documentation has always been on the project critical path and usually is the last item completed. I've seen this delay more projects than I care to mention. With you as the only reviewer, I anticipate that you'll be the bottleneck in the entire program."

Sue answered, "Dan, if I need help, the Quality group will provide that help. Let's not look for problems now. Let me finish."

Dan kept on talking, "I disagree! Our Six Sigma design policy specifies that these types of issues must be identified early in the program. I would like to see additional names from your organization listed in

<u>Project Charter</u>

PROGRAM TITLE:

Draft or Final                                                          Initial Approval Date:

Date Finalized:

| Project Team: | Primary<br><br>Name/Phone # | Backup<br><br>Name/Phone # | Director Level Manager Approval |
|---|---|---|---|
| Customer | | | |
| Program Manager | | | |
| Project Engineers:<br><br>• Process development and equipment design<br><br>• Inhaler design<br><br>• Inhaler injection molding<br><br>• Packaging | | | |
| Quality | | | |
| Medical Affairs | | | |
| Marketing | | | |
| Materials Management | | | |
| Plant Operations | | | |

| Objectives: |
|---|
| Design and develop an inhaler system to deliver two dosages of the Oxy-Fox cocktail. |
| Develop product medical insert. |

| Definitions of when program is complete |
|---|
| 1. Process Qualifications |
| 2. Verification/Validation |
| 3. Oxy-Fox Inhaler launch |

**Figure 4.1**   Project charter.

| Main Deliverables | |
|---|---|
| Deliverable | Description |
|  |  |
|  |  |
|  |  |

| Key Assumptions and Necessary Conditions: |
|---|
|  |
|  |
|  |
|  |
|  |

| Timeline/Schedule: | | | | |
|---|---|---|---|---|
|  | Milestone Description/ Dates: | | | Buffered End Date: |
| Major Project Milestones | Plan Date | Latest Best Estimate | Completion Date |  |
|  |  |  |  |  |
|  |  |  |  |  |
| Complete Project Charter Approvals |  |  |  |  |
| Design Review - Inputs |  |  |  |  |
| Design Review - Outputs |  |  |  |  |
| Verification/Validation Design Review |  |  |  |  |
| Risks: |  |  |  |  |
| Risk Description | Risk Owner | Plan to Address | | |
|  |  |  | | |
|  |  |  | | |

**Figure 4.1**    (*Continued*)

| Documentation: | | * Reference specific design requirement. | |
|---|---|---|---|
| Document #/Name | * | Person Responsible | Description |
| 1. Process specifications | | | |
| 2. Qualifications | | | |
| 3. Inhaler device sign-off | | | |
| 4. Final product Drawings | | | |

| Documentation Key Assumptions: | |
|---|---|
| | |
| | |

| Documentation Timeline/Schedule | | |
|---|---|---|
| Document #/Name | Completion Dates | Responsibility |
| | | |
| | | |
| | | |

| Documentation Risks | Risk Description, Owner, and Plan to Address. | |
|---|---|---|
| Document #/Name | Risk Description / Plan to Address | Risk Owner |
| | | |
| | | |

| Project Charter Approvals | Date Approved |
|---|---|
| Program Manager | |
| Program Customer/Sponsor | |
| Pharma VP Operations | |
| Pharma VP Medical Affairs | |
| Pharma VP Quality and Compliance | |

**Figure 4.1**   (*Continued*)

the charter now—not wait until the end of the project when it will be tough to get help."

"Dan, your point is a good one and I'll take it to my boss. I'm sure this can be resolved. Is it okay if I continue? Any more concerns from the group? Keith . . . any concerns?"

Keith answered, "Not at this time but this was supposed to be an introductory meeting and we're running out of time. The meeting room is only booked for two hours and Gonzales already blew the schedule."

Dan Garvey raised his hand, "Sue, I have a question. What is the difference between the verification and validation plans?"

Sue answered, "The Verification Plan is a preapproved protocol including rigorous testing that the actual device conforms to the intended product design. The Validation plan is also a preapproved testing document demonstrating the design meets customer requirements.

"The most important part of the charter is the final approval by executive Pharma management. Management endorsement assures the team we will get both financial and resource support to succeed. Following the project charter a detailed description of the 'Customer Requirements' must be completed and approved by the team members, Marketing, and Medical Affairs. The Customer Requirement for our stage of this program may not have to address the efficacy of the drug— this will be done by Medical Affairs, but it must address the requirements for the inhaler and the Oxy-Fox process. A simple example is the size of the inhaler, the quantity of drug that must be metered, how many dosages, etcetera. More specifically, the Customer Requirements, or in Six Sigma terms the 'Voice of the Customer,' or in Agency terms the 'Design Input Document,' which defines the product performance, safety, reliability characteristics, environmental limits, physical attributes, compatibility with other devices, applicable standards, Agency requirements, packaging specifications, and labeling. Basically, this document addresses the intended use of the inhaler to the extent of what color it should be, the ergonomic design such as how the inhaler is held, and so on."

Keith interrupted, "Excuse me Sue—I'll have a Marketing rep on the team for the next meeting. Sorry to interrupt"

Sue thanked Keith and continued, "The 'Design Inputs' must be mapped to the actual design. The actual design in Agency terms is called the Design Outputs. The Design Outputs are the specifications, drawings, process specifications to produce the inhaler and Oxy-Fox cocktail, and all support documentation for the device. Once the preliminary design is determined, a 'failure mode and effects analysis,' which you will recognize as an FMEA, at the product-use level, will be done by the design team and Quality. And yes, we absolutely have to have Marketing involved. A 'Traceability' document is also required linking the design outputs to the design inputs with all appropriate testing noted.

"Next we need a 'Technical Feasibility Document,' which outlines in detail the design and process requirements for both the inhaler and the manufacturing process. The feasibility document is an internal Six Sigma requirement. This will require significant product development and engineering input. I expect that prototypes will be made for the inhaler as well as a bench-level process developed to establish process parameters. And, as Dr. Gonzales pointed out a number of times, inhalers produced from these prototypes can be used for the clinical studies as long as the processes are validated, and that includes the injection molding of the inhaler components. Just make sure that objective evidence is eventually provided in the DMR that the final manufacturing processes are equivalent to the validated prototype.

"Once this document is reviewed and approved we enter the Design Control phase. This phase continues for the remainder of the project and will require any changes to be documented with objective evidence of why the change was made. Any changes to the prototypes must be documented as well. This must be closely managed by Gordon and Dan, and it must be reviewed and approved by me prior to the changes being incorporated in either the inhaler components or the Oxy-Fox process."

Sue paused, looking at both Gordon and Dan to make sure they understood, and continued, "Design Reviews by management, 'Tollgates' in Six Sigma terms, will be periodically scheduled after the completion of each of the document phases. These are important . . . and must be done.

"Once the design of the inhaler is complete and the process parameters are set, a design FMEA will be conducted by the team. The team will review each component of the inhaler and determine what can potentially go wrong and the effect of the failure at both the device and patient levels. An example would be if the metering component of the inhaler malfunctioned—what effect would the malfunction have on the device and would this cause injury or death to the patient. The FMEA would then address if there were modes of control in place to prevent the occurrence or if modes must be added to the design and/or manufacturing process to keep this malfunction from occurring. A typical mode of control could include 100 percent inspection if that inspection was nondestructive.

"Finally, the team will develop verification and validation plans to test whether the inhaler and the process meet the design outputs and whether the actual product meets the customer requirement, respectively. More specifically, the Verification Plan describes the testing, objective evaluations, inspections, and acceptance criteria to ensure that the defined processes are being followed correctly and that the product specifications can be met. As I mentioned, the Validation Plan is designed to ensure that the customer requirements are met. This plan must be performed under actual operating conditions with the final equipment used for the initial market entry. To repeat myself, bench-type equipment can be used to support clinical and initial market entry as long as the equipment is validated and the final equipment is shown to be equivalent to the clinical equipment by objective evidence. Each plan must include multiple production lots of both raw material and finished goods to test for lot-to-lot material variations and lot-to-lot manufacturing variation—a minimum of three lots will be required to satisfy these requirements. If possible, I'll allow the same material lots to be used in both qualifications. Both plans must be completed and approved prior to any actual acceptance testing and will include statistically determined acceptance levels demonstrating that the processes are under control and are capable of meeting process parameters and product characteristics. Results from the in-house equipment qualifications will be referenced as objective evidence in meeting the requirements set forth in both the Verification and Validation documents. Equipment used to produce components in the inhaler produced at outside suppliers, specifically the injection-molding houses, will be qualified and referenced in the DMR but the actual packages will be kept at the supplier. I volunteer to work with the Statistical Methods group in my organization to establish the acceptance criteria, product sampling procedures, and quantities for each lot needed to establish statistical significance. Medical Affairs will have input in defining the patient risk levels for the acceptance criteria."

Sue paused, taking a breath, "In addition, I'll work with Dave Stall to establish a Purchasing Control Document to test all incoming raw materials. Dan, I expect you to establish qualification and quality acceptance plans with the molding suppliers for the inhaler plastic components."

At this point Dave Stall raised his hand and asked permission to speak. Dave had been instructed by his boss that the Materials Management group would handle all molding supplier communications and management of their activities.

"I volunteer to do this. Dan has more than enough work with the Oxy-Fox process development."

Dan responded without raising his hand, "Not on your life. The molding suppliers have dealt with my group for the last ten years and will continue to do so. This is a technical evaluation and the engineers in my organization will work with these guys in the detailed design of the plastic components, the design of the molding process, and the fabrication of the molds. Your group is not technical . . . this is not a K-Mart product and can't be handled by a simple purchase order. You forget that there're thirty-eight engineers in my group. I manage this function—I don't do the work myself. You can handle the supplier contract negotiations, which will remain under my direction, and establish the pipeline quantities needed for market launch, just like any other product introduction. The identification of multiple raw material sources will give your group more than enough to do. If there's a problem, have Jennifer Feddler talk to me."

Keith Carlisle stopped Dan immediately. "Dan, no need to get upset. This will work out. Nothing will be changed that could jeopardize the success of this program."

*Dan knows that Jennifer will come nowhere near him and will try to have the Operations VP, Larry Fletcher, intervene. But hopefully Larry will only give her lip service . . . the molding group is considered a strategic initiative in the success of product launches and continuation of the business. Dan has been told by people at the corporate level that Larry holds a very high opinion of his technical capabilities. Though Larry has never liked Dan's management style . . . because Dan is outspoken and has openly disagreed with Larry in a number of situations when he should have just kept his mouth shut and gone along with the program. Larry has tried to move Dan out of the Engineering Group Manager role to the level of an individual contributor with no management responsibility without any success . . . Dan is too well respected by the research organization and he has the Director of Engineering's, Frank Hamilton, full support in his management role, not to mention the influence of his buddy Janet Weatherbe, Corporate Vice President of Human Resources. Larry has had his Human Resources Director for Plant Operations, Ralph DeSousa, monitor Dan's associations with his reports and people in other organizations with the intent of finding an angle to get him out of the Group Manager position but with even less success. Dan is smart and this will never happen. Dan is no friend of Larry. Dan is also aware that Keith Carlisle is envious of his influence with the R&D organization and has aligned himself with Jennifer to move the molding business under her direction. Jennifer plays company politics better than anyone and could be an up-and-coming force in Kinnen. Keith Carlisle will do whatever he must to be on her good side. Time will tell.*

Keith Carlisle continued, "Sue, we are almost out of time. Can you summarize the rest of your presentation? We're not going to have time for Product Development or Engineering to go over their plans."

Sue continues, "OK, almost done anyway. The team mustn't forget that traceability of all supplier lots received for initial investigation, testing, and . . . even more important, any material lots used in clinical studies and the actual verification and validation . . . must be strictly maintained. This is major. And the finished devices cannot be released for market distribution until all activities identified in the DMR are complete—the documents are reviewed by the Quality group, the release is authorized by the management signatures, and the authorization is dated. My final point is a question.

"Keith, will the first lots to stock for initial market launch be produced at the corporate plant?"

"Yes. And once the organization has the initial pipeline built, production will be moved to the offshore plants. Only the product from Argentina will come back for domestic distribution. Both are longer term production strategies."

Sue shook her head, saying "I figured as much. I'm sure the team is aware that we've had issues with Agency cGMP audits at our Argentina facility involving inadequate design transfer documents and in-process finished-goods separation and labeling. We haven't received a warning letter from the Agency but corporate-level Quality management is concerned that a warning letter will be sent sooner than later . . . as an organization we have done nothing to improve our product transfer qualifications. As part of the DMR, I want a complete transfer plan to include the same verification and validation qualification plans at the Argentina facility as done for product launch at our site. This definitely includes equipment qualifications and any plastic mold transfers. I'm not worried about the Euro transfer; the staff at this plant is more than conscientious and has always followed the transfer plan guidelines. And the Euro product will not be sold domestically and will not be under Agency audit requirements.

"One more thought is product labeling and the insert. I'll set up meetings with Medical Affairs to start this rolling. Keith, that's all I have for the time being."

"Thanks Sue. I think the team has gotten the point. There's another group outside the door waiting for the room. I'll schedule another meeting to finish this. Have a good day!

"Gordon, Dan . . . I need you two to get together and finalize your development plans prior to the next meeting. And I want your best

scenarios reflected in these plans—no sandbagging. Is this clear?"
Neither Gordon nor Dan answered.

\*\*\*

Leaving the meeting Gordon asked Dan what he was doing for lunch—
he needed advice. The two lunched in the Kinnen cafeteria. Dan was
apprehensive; Gordon was a private guy who never asked for advice.
The daily special consisted of country fried chicken with mashed pota-
toes covered with pepper gravy and a side of buttered corn—Dan's
favorite—and Gordon picked up the check.

Dan asked, "What's up?"

"I'm tired of the continual abuse in my one-on-ones with Chase.
And Chase does his best to humiliate me at every staff meeting. I'm
at the point where I need to do something. Dan, we've worked
together for a long time and I know you wouldn't stand for this. What
should I do?"

"You have to stand up for yourself. Don't let him do this. Chase is a
predator and senses you'll take all the abuse he can put out. Stop taking
it. Stand up to this guy."

Gordon answered, "Easy for you to say. You don't report to him."

"Gordon, I'll talk to Janet Weatherbe and bring this to her attention.
I know she's aware of his abusive personality and from a Human
Resources perspective she is tough. When she was in Pharma we hung
out. We're friends. But you have to take control. I met Chase at a joint
Pharma–Therapeutic engineering meeting when he was Engineering
Director of Therapeutics. He presented the Therapeutic Engineering
Plan for the year and I questioned why he needed the budget presented
for just maintaining the business. He totally flipped out . . . yelled at me
that it was none of my concern and actually turned blue. At lunch I
asked Jon Flanen, an engineer in his group whom I've known for some
time, what was up with this guy. Jon told me Chase has a serious high
blood pressure problem as well as he's a complete ass. Jon went on to
say at one meeting he presented a project that was running over budget
and Chase screamed at him . . . threatening to have him fired, and actually got up and punched the wall. He's mentally unstable . . . he's always
been that way. Use this to your advantage. When he openly degrades
you go back at him. Go on the attack . . . don't just sit there and take
it. Push him as far as you can . . . he may have a stroke."

Gordon didn't say anything. He just looked at Garvey, thinking that
he could never do it.

*Dan knew he should stay out of it. Gordon was a big boy and had to learn to take care of himself, but he would talk to Janet and bring it to her attention. This was a human resources matter and Janet was Corporate VP. A conflict would only hurt the relationship between Product Development and Engineering in the long run ... it had to be resolved.*

*Dan was confident that Janet would get this resolved. Prior to her promotion to the Corporate VP Human Resources position she was in Pharma and created a number of programs, including educational opportunities for the hourly employees and promoting additional benefit packages for the division, and she established meaningful evaluation standards based on achieved skill sets for scientific and engineering promotions and pay increases.*

*Dan and Janet have worked together for a number of years and are good friends. Janet is where Dan's power lies.*


## 4.2   MEETING WITH MEDICAL AFFAIRS: TOXICITY STUDIES

Late in the day Keith Carlisle meet with Dr. Gonzales and Susan Vickory MD. Dr. Vickory is Corporate Vice President of Medical Affairs reporting to the Pharma President. She was recruited from a world-class medical school, where she was Professor of Medicine, Department of Cardiology, and has been in her current position for about five years.

The meeting was scheduled at the last minute and was short. Dr. Vickory said she had evaluated the due-diligence report and Katlin data and found no objective evidence that the required toxicity studies were done. But this may not be a big deal. Both oxytocin and tricyclics have been used in previous animal and human studies and are well characterized in the literature. Tricyclic compounds have been administered intravenously in human clinical settings and the nasal spray would have a reduced bioavailability in blood circulation. The tricyclic literature studies will be incorporated into the IND documents and submitted to the Agency. She is confident from discussion with Mark Brooks, Kinnen Chief Executive Officer, that the Agency will accept the literature studies even though there is no objective evidence of administration by nasal spray. In addition, it is Mr. Brooks' opinion that since both oxytocin and tricyclics are well characterized the Agency will approve an abbreviated human study plan. The abbreviated testing should allow the program to be completed in no more than two years and possibly less.

*Dr. Vickory had personal reservations on the efficacy of the foxepin compound itself but would keep this to herself. She reviewed the university studies and noted foxepin was not a classical tricyclic compound and affects the dopamine pathway quite differently—and the animal toxicity studies were done by oral administration, not intravenously. She'll not go against Mr. Brooks on this decision, and the successful launch of this product means big bonus money.*

At the conclusion of the meeting Dr. Vickory asked Keith Carlisle to stay a few minutes and excused Dr. Gonzales.

Dr. Vickory started the conversation, "Keith, I know how important the launch of this new product is to the performance of our division and plan not to be the show stopper. I'm going to share information with you from the university development of the foxepin used in the inhaler. You can make the decision whether this needs to be brought to the team's attention or not. The university studies were based upon oral administration of foxepin, not intravascular or intramuscular injection. In an orally administered drug, the drug is metabolized in the gastrointestinal system and the liver. Normally the potency of the drug through oral administration is reduced before entering the body's blood circulation. This is a problem in the due-diligence report. By intravenous injection there's no reduction in potency and 100 percent of the drug could pass to the systemic circulation; in the case of this type compound the drug passes readily through the blood–brain barrier. I'll ignore this because the NDA trials will be done with human populations with dosage ranges that should identify a problem if one exists. We'll have to take a risk that there will be no complications in the results of the clinical studies."

Keith asked, "Is Gonzales aware of this."

"No. He's only reviewed the Katlin animal studies. The animal studies did administer the foxepin intravenously with no aberrant reactions but toxicity over a range of dosage was not investigated."

*Keith left the meeting a happy man . . . his career plan was back on track and wouldn't follow the normal course. Keith decided not to bring the information from Dr. Vickory to the attention of the team. After all, the clinicals will test the foxepin cocktail in human studies. Larry has expressed to Keith, with no misunderstanding, that his management bonus will be significantly increased if he meets a product launch date of less than two years . . . and hinted at a reorganization that would include a new position of Director Program Management. Keith will push the prototype process developments so that the clinical studies can go concurrently with the prototype validations. Keith's lifestyle relies on this bonus.*

# 5

# MEETING MINUTES GUIDELINES

## 5.1   LAUNCH TEAM MEETING NUMBER 2

**In Attendance**

Keith Carlisle, Program Manager

Susan Jaffey, Quality

Dan Garvey, Engineering

Gordon Taylor, Product Development

Dave Stall, Materials Management

Patty Keyser, New Product Marketing Manager—reports to Maria Sanchez, Pharma Director Marketing and Sales

**Not in Attendance**

Dr. Gonzales, Medical Affairs

Keith Carlisle tried to set up a second meeting for the end of the week but schedule conflicts of the key participants pushed the meeting to week 3 of the program. Keith does not like to lose time—time is the enemy of the project launch team.

*A History of a cGMP Medical Event Investigation*, First Edition. Michael A. Brown.
© 2013 John Wiley & Sons, Inc. Published 2013 by John Wiley & Sons, Inc.

Keith is an experienced program manager and as a Six Sigma Black Belt follows a prescribed meeting format to the letter. Keith publishes an agenda prior to the meeting—sent to each team member with a copy to their management for information, as well as to all stakeholders in the organization who require knowledge of the team's progress. The agenda lists action items determined from the previous meeting and the person responsible; it lists discussion items for the upcoming meeting; and it identifies the meeting's expected results.

Following the meeting Keith usually publishes minutes in a timely manner, no more than one day after the meeting unless there's a politically sensitive issue that must be addressed with management before the minutes are issued. When out-of-the-ordinary situations arise, upper management must always be informed prior to the stakeholders copied on the meeting circulation list.

In addition to the weekly meeting minutes Keith publishes a monthly report covering all pertinent aspects of the program, including timing and budget, team progress in each area, and all significant decisions; this report is circulated to all management and other stakeholders. A separate report is published covering the status of all required documentation, including product labeling.

Meeting minutes address the following:

- Meeting date, location, and time
- Team members in attendance and their function
- Action items from the previous meeting that needed to be addressed, the person responsible, and resolution of the action item
- Agenda items for this meeting and a summary of the discussion
- Identification of all decisions and action items, with assigned responsibility and timing
- Any potential issues that impact program timing or budget, identified as a separate item
- Action items that will be addressed in the next meeting, with responsibility
- Location and time for the next meeting

Keith tries to manage the team meetings so that each member is encouraged to participate in the discussions: he tries to resolve conflict between the participants; he tries to participate as a problem solver; and he tries his best not to be part of the problem. Most important, Keith tries to be a leader. Keith is not always successful in this role.

There were no unresolved items from the last meeting so Keith had an abbreviated agenda covering the development plans from Engineering and Product Development, a market entry review from Patty Keyser, material sourcing, and product qualifications.

Dan Garvey began, "Keith, I need the chemical formulations from R&D and the preliminary inhaler design before a detailed engineering plan can be established. I'm working with Jeffrey Daniels of R&D to get this done. Jeff is aware of how critical this is and it is his first priority. Gordon needs another month before any preliminary device molded part specifications and drawings can be developed. With these two items my group can start a bench-level process and begin working with the molding supplier on the actual design of the plastic components. Initially prototype four-cavity molds, also called tools, will be built to provide samples for characterization studies. Once the process parameters are finalized from the bench-level prototype and actual inhaler components are produced and validated from the molds and are assembled, Medical Affairs can begin the clinical studies. I want to take this time to review the . . ."

Keith interrupted Dan in the middle of the sentence. "Not acceptable, you've had more than a week to get the chemistry information from R&D. I understand that Gordon needs time for the inhaler design but chemistries should be known. I mean, we've had the Oxy-Fox cocktail in-house for more than a month."

Dan answered, "I'll talk to Jeff after the meeting. Let me finish and try not to interrupt."

*It is obvious that Dan is not going to let Keith manage his responsibilities. He doesn't trust Keith. Dan feels Keith has his own agenda based on his personal success goals.*

Dan continued, "I want to review the cGMP and Six Sigma requirements that my group will follow for the Oxy-Fox process . . ."

Again Dan was cut off in the middle of his sentence following a knock on the door. Dan's administrative assistant stuck her head in the room telling Dan that his wife was on the phone and it was an emergency. She needed to talk to him immediately.

Dan started to leave the room but Keith Carlisle objected, telling him that he could take care of this later or have his admin do it. Dan ignored him and left.

\*\*\*

As Dan walked down the hallway—his office was on the same floor as the meeting room, about a hundred-yard walk, he thought to himself,

*"The last time Becky called and needed me immediately was when our six-year-old son didn't show up for school. This was a real emergency and she hasn't called me since. She doesn't call just to say hi or ask me how my day is going . . . Michael just took a roundabout way to get to school and got lost wandering around the high school parking lot six blocks west of the grammar school. The police found him and brought him to school but didn't inform Becky at that time. . . . Becky does not call!"*

The admin transferred the call and Dan picked up the phone in his office.

Anticipating the worst, Dan answered "Becky, what's up?"

"There is an equipment problem on filling line number 6. Bottles are jamming and maintenance doesn't know what to do."

Realizing that this was not his wife, Dan asked "Who is this?"

"Becky . . . Becky Seimans." *Becky was obviously annoyed for not being recognized immediately. Dan knew that she had responsibility for product availability—the fill line being down could delay shipment dates and the plant maintenance group should take this as their first priority: not his equipment group.* "Dan, you need to get out to this line and fix the problem."

"Becky, I don't appreciate being pulled out of a meeting for this. Contact Plant Maintenance and request their engineering support. I need to get back to my meeting and am hanging up."

Dan hung the phone up in the middle of Becky Seimans' protest.

As Dan walked the hundred yards back to the conference room he heard a loud, pounding sound as if someone was running. Becky's office was just on the second floor of this building and, as he turned, she was coming after him. Becky is a rather large woman—almost six feet tall with weight to match—about twice the size of Dan. Dan hurried to get to the conference room and, as he got closer, Becky got closer to Dan. It was a race to safety. As Dan entered the meeting, he kicked the door shut behind him. The door hit Becky squarely on the head. She did not fall but retreated from her attack on Dan.

With the door closed Dan started the presentation again, "Okay, the emergency was no big deal. Let's continue with where I left off. No more interruptions.

"As I was saying, since I don't have adequate information to establish a development plan, I want to review the content so both Gordon and Sue can get an idea of what is required . . ."

The door to the meeting room opened again and his admin told him that his attendance in the Engineering Director's office was required.

The Director was two offices away from Dan's office and he walked the hundred-yard hallway again.

*Frank Hamilton is Pharma Director of Engineering— responsible for all engineering functions for the division, including facility design and construction, engineering project management, product packaging, and engineering support services, which Dan manages. Frank delegates authority to his group managers—he does not believe in micromanagement unless there is reason to do so. He manages all department progress through weekly one-on-one meetings with his direct reports, staff meetings, and through project update presentations. Frank supports the engineers politically and is respected across the corporation as a true leader, and he chairs the Kinnen Engineering Council. His motivation is the continuous success of the Engineering Department's responsibilities. Frank has direct responsibility for all engineering aspects of product development and initial product launch, which he delegates to Dan. Frank is on a fast-track program for upward movement within the Kinnen corporation.*

Dan knocked on Frank's door and was told to come into the office. The office was furnished comfortably; it had a personal work area with windows on two sides overlooking the building's main entry, and a cluster of five chairs surrounding a round table. Dan has attended many meetings in this room and the office was designed so the participants would be at ease. Dan has reported to Frank for over six years and the two have a deep respect for each other. Dan sensed that Frank was upset.

Frank started, "Dan, Becky Seimans just left my office. I can't believe you actually hung up on her and hit her with the door. She's extremely angry and going to formally complain to Human Resources. What the hell is going through your mind?"

"Frank, she had no business calling me out of a program meeting to complain about problems on the fill lines. There's a procedure in place requiring her to contact plant engineering for these kinds of things. I told her that I was in a meeting and was hanging up. I *told* her that I was hanging up. If she persisted in the conversation, there's nothing I can do about it."

"But you hit her in the head with the door. I want a full written apology within the hour and this will go into your personnel record."

"Frank, I was scared. I didn't mean to hit her with the door but she was right behind me...running. You see the size of this woman! She was actually chasing me...running at full speed. I really was scared and I want an apology from her. No way am I writing anything. If she wants Human Resources involved, I have a complaint of my own."

Frank told Dan in simple terms that he had to understand, "We're walking on eggshells with her organization as it is. Jennifer is constantly politicking to take over our molding business. This relationship isn't going to get us anywhere. You're a group manager, you have two master's degrees, and you're smarter than this. You have to foster these relationships a whole lot better than you are doing. We are a Six Sigma company . . . read your manual."

"Not my problem; you need to deal with this at the director level. I don't have time for division politics. Jennifer will have you reporting to her if you're not careful. I need to get back to this meeting or you'll have Keith Carlisle in your office next."

"Okay, go. Forget about the apology."

"Already have." He was thinking, "*Guess I'll be attending the next 'Building Positive Work Relationships' seminar . . . again.*"

Dan walked the hundred yards back down the hallway to the conference room for who knows how many times that day. When Dan entered, the room was empty with only Keith Carlisle still there.

"Dan, I don't appreciate this."

"Not my fault. Schedule a follow-up meeting. I'll meet with Jeff this afternoon and work out the chemistries. By the way, Jeff should be on the project team."

# 6

# PROJECT TIMING, MARKETING PLAN, AND OFFSHORE MOLDING

## 6.1  LAUNCH TEAM MEETING NUMBER 3

**In Attendance**

Keith Carlisle, Program Manager

Susan Jaffey, Quality

Dan Garvey, Engineering

Gordon Taylor, Product Development

Dave Stall, Materials Management

Patty Keyser, Marketing

Jeffrey Daniels PhD, Research and Development—staff scientist responsible for the development of the Oxy-Fox drug

**Not in Attendance**

Dr. Gonzales, Medical Affairs

Keith again presented an abbreviated agenda covering the development plans from Engineering and Product Development, material sourcing, product qualifications, and a market entry review from Patty Keyser.

*A History of a cGMP Medical Event Investigation*, First Edition. Michael A. Brown.
© 2013 John Wiley & Sons, Inc. Published 2013 by John Wiley & Sons, Inc.

Dan began, "Sorry for the interruptions in the last meeting. Jeff and I met and have the chemistries for the Oxy-Fox cocktail. The formulation is a combination of the hormone oxytocin and the tricyclic foxepin. There'll be two versions of the cocktail—each version will contain 25 milligrams of oxytocin with different dosages of foxepin: 25 milligrams and 50 milligrams. The concentration tolerances will not be an issue in the process formulation. What will be an issue is that the two drugs are totally different: foxepin behaves much like a hydrophobic compound, and oxytocin is a hydrophilic peptide. They do not mix—in fact, the two are like oil and water. R&D is looking into a carrier compound to provide a binding substrate. My group started the design of a benchtop conjugation and mixing process late last week and we expect to have a prototype complete in two weeks but will not be able to go any further until the carrier is available. Jeff, anything you can add?"

"As a matter of fact there is a chemically nonreactive binder approved for human use by U.S. Pharmacopeia. The carrier is based on a complex with both hydrophobic and hydrophilic ends—much like soap with a hydrophobic head and a hydrophilic tail—and will bind both drugs as a single molecule. I'll work with Dan's group in establishing the conjugation process parameters. The substrate is on-order and will be in-house by the end of the week. Even with the two molecules conjugated, patient administration will still have to be by nasal spray. The peptide bound to the molecule will be destroyed by the gut micro-organisms. If there are no questions, I'll turn the meeting back to Dan."

Dan continued, "On a parallel path we're investigating a system from Haul-Miller designed to dispense the cocktail into the container and install a special cap required to interface with the nasal pump. The container will have to be an amber bottle due to the light sensitivity of the conjugated molecule, and the cap must be hermetically sealed aluminum capable of holding the pumping mechanism. Haul-Miller has a benchtop system that is an off-the-shelf item that can be delivered in two weeks from receipt of purchase order. I instructed Purchasing this was a priority and they assured me the order would be faxed by the end of day tomorrow. We need to characterize the dispensing parameters required to dispense the cocktail into the container and the method to install the aluminum top. The pumping mechanism will fit through an opening in the top. We will also have to quantify the amount of the drug the inhaler will administer in each patient application based upon ambient temperature and pressure. This will be the most difficult part of the development as it involves the inhaler component design as well as complete characterization of temperature and pressure limits to keep the cocktail from breaking down to its constituent compounds.

I can give you a rough schedule now and a complete Microsoft project schedule by the end of the week. Keith, the schedule must be treated as preliminary at this point but it is good enough to go into the draft Project Charter. I'll include a minimum amount of time to account for three revisions of the prototype at two weeks per revision for a total of six weeks. So, you can figure that we'll need about sixteen weeks to complete the prototype, develop the conjugation process, run a chromatography study to determine molecular weights and separation temperatures, receive the Haul-Miller dispensing and capping equipment, and establish the process parameters and validation plan. Take the beginning date as Monday next week. Once Gordon has completed the preliminary design of the nasal device, my group can start the detailed component design and prototype mold fabrication with the molding suppliers. At this time I can't give you timing until Gordon and I work this out."

Dan paused to gauge if Keith Carlisle has objections, then continued, "To make sure that the final manufacturing process is equivalent to our prototype, the engineers will do a complete equipment process qualification. The mold prototypes will be four-cavity hot-runner tools to account for deviations in plastic flow—and will be qualified as well. The output from these molds will be used in the inhalers for clinical studies and market launch, providing pipeline material. Once marketing has established a long-range market plan, the molding process will be scaled up to support their plan. This scale-up could take up to a year or longer depending on the product quantity detailed in the long-range plan."

Keith was obviously not happy with this timing and put Dan on the spot, "Is there anything you can do to reduce the prototype timing, say, to ten weeks. Also there is no requirement that either the prototype process or the molds have to go through the qualification process prior to releasing the inhalers to the clinical investigators."

Sue joined in, "It's an excellent idea to do these qualifications and the Quality group will provide complete support. In fact, the Agency requires complete validations. And Keith, Dr. Gonzales pointed this out."

Dan anticipated this response from Keith and said, "Keith, Sue is correct. And this is my best timing—no sandbagging per your request. The schedule will be a stretch as it is. Also the timing is just for the prototype equipment. The final production process can't be defined or sent out for bid until completion of the prototype and verification of the process parameters. And assembly equipment will be needed for the inhaler. At this time figure twelve to sixteen months for final manufacturing equipment delivery and process qualifications if the total equipment program is done in-house."

Dan paused for a moment, wondering whether he should mention his plan to potentially reduce the overall timing of the final assembly equipment; he decided what could it hurt. "During the prototype phase we'll be working with an outside supplier to provide a production line for a finished device. That is, the conjugation and mixing of the Oxy-Fox cocktail, as well as container dispensing, will remain in-house but the final assembly of the inhaler components could be farmed out. This could reduce the overall project schedule substantially."

Dan paused again to make sure Keith Carlisle was listening, keeping eye contact, while he said, "But this idea is way too early to advertise to upper management. Once Larry gets his hands on this, it'll be locked in stone. Not sure if we can even do it, so Keith, keep it to yourself."

Keith responded, "I don't accept the timing as you've presented it for the prototype equipment and I want your group to take a better look at reducing the schedule. Make the move of the final assembly equipment to an outside supplier a top priority—reducing the overall schedule will make the entire team look good in Larry's eyes . . .Gordon you're next on the agenda."

Gordon took the floor, "Dan has pretty much covered what product development has to do over the next four to six weeks. I was part of the due-diligence team working with Katlin in transferring the technology to us. We negotiated a license agreement on the foxepin compound and their design of a nasal inhaler that was quite effective—Bob Jameson, our Financial VP, wouldn't accept the agreement to include the inhaler. In his words, 'We'll not pay that much for something we can design ourselves. This is not the only inhaler . . . our development guys can do this for a third the cost.' To this end, we've a number of devices from competitors that are being reverse engineered at this time. The patent attorneys are looking into each device and will give us an opinion by the end of the week if we can copy or modify without infringement. This will save a few weeks in development since we don't have to start with a blank sheet. My group will work with Engineering on the final details. All components of the inhaler will be designed on a solid model CAD system compatible with the molding supplier Dan has chosen to do the final design. Our designs can be electronically transmitted direct to the supplier to save time. From these preliminaries, final designs incorporating the plastic material properties will be developed and used to construct the molds. This has been done before and is the most efficient method."

Keith said in somewhat sarcastic voice, "Okay. From Dan's expression I won't even ask you if your group can do this sooner."

Keith continued, introducing the marketing representative. "Most of you do not know Patty Keyser from marketing. She'll review the marketing plan and also has some ideas on how the inhaler should look from a woman's perspective. Patty you have the floor."

Patty began, "Thank you Keith. With all the talk on schedules, processes, and the inhaler, I want to tell you what we want to do in the design. Marketing has hired a design firm to develop the esthetics of the device suitable to a woman. All you men doing the development scares me. We haven't met with this firm yet so I can't give you any timing.

Dan Garvey immediately spoke up, "Patty, I respect marketing's position on a design that would be attractive to women. My group focuses on the technical design of the process and Gordon focuses on the overall design of the inhaler. You have complete freedom in picking any surface texture, shape, or color your design firm determines acceptable to a female user. My only condition is that the design cannot interfere with or require any changes in the technical function of the inhaler. And your firm has to work to our overall project schedule. We can get together off-line to go over this further. I'll set up a meeting to include Gordon. Can you get a representative from the design firm to attend?"

Patty answered, "Sounds good, I'll try. The design contract is still under negotiation and I'm not sure of their availability. The team should also know that the design firm will be conducting at least two user focus groups—one in the Boston area and one in San Francisco. This could take time. I would not want to commit to a design until a final report is accepted from marketing management."

At this point Keith Carlisle should have interrupted Patty and told her that the program was on an internal fast track and the plan did not include the studies she had in mind. But it was a political issue and he would handle it at the Larry Fletcher level.

Seeing no objections to her talk so far, Patty continued, "Marketing has not really had a lot of warning on this launch and we're just getting our ducks in order. Initially, I can say that the main market for this product is domestic. I've been told that we would have the only product available specifically for postpartum depression and this potentially could be a billion-dollar business. We're working with our sales offices around the country to survey physician offices to try to get a better quantification of the extent of the market as well as establish a market launch plan. I'm confident that we'll only enter one geographic sector of the country initially. One market would be about as much as the sales and customer support groups could handle."

Realizing that Patty was done Keith Carlisle said, "Patty, good review. But we'll need initial market quantities as soon as your group can get them. Next on the agenda is an update from Materials Management. Dave Stall, you have the floor."

"Before I get into any material plans I want to ask why a molding supplier has been selected so soon into the program. All bid documents have to come through my office and I haven't seen any. It's company policy that all molds go out for bid. I'm very concerned that choosing a supplier at this point will lock us into a long-term commitment. Jennifer has mentioned that if there is a large quantity of molded parts we'll take this offshore, possibly China."

At this point Dan was ready to freak out. "Dave, we must have a molding company onboard to consult on the preliminary component designs, complete the final designs, and fabricate the molds for the clinical studies and market launch if we're to meet the schedule. If we bid this out, add an additional ten weeks to the program before any parts come to us. This is how all our programs have gone in the past and this is what we'll do now. I have authority to make this decision. Once we start the mold build program to provide components for the long-term market strategy, bid packages will be sent out. And I do not even want to hear about offshore molding . . . and China . . . does your group live in a cave? Every day companies who sent their molding operations to China are bringing them back to the States. And I will not even go over the quality histories of some China imports."

Dave Stall, maintaining his composure, answered, "We can have a molding bid ready in no more than two weeks. By just giving this work out you're giving this supplier an unfair advantage when the long-term molds are bid."

Dan responded, "Dave, two weeks to have the bid is ridiculous . . . there's no information to even put into a bid package at this time, and add another six weeks to evaluate the bids and choose a competent supplier. With the schedule we're working toward . . ."

Keith Carlisle stopped Dan, "Dan, that's enough. Dave get off this track with the molding operation. Your two organizations work this out off-line; this is not the place. Go on with the material plan."

Dave Stall was obviously upset but continued, "Once the compounds are completely characterized by the development teams, Materials Management will investigate a sourcing strategy. At this time the sourcing strategy is not on the critical path but I'd like the team to source the development compounds from suppliers that are on our preferred supplier list. That's it for me."

Jeff spoke up that his group had the list and would follow Dave's request.

Keith Carlisle continued, "We've run out of time; actually we're five minutes over. Another group is in the hall waiting for us to get out of the room. I think a lot has been accomplished in today's meeting and we'll review the qualification plans next week. Dan, you have an assignment to reduce your timing. We can get together off-line to review before the next team meeting. Minutes will be published by the end of tomorrow and I would appreciate you all reviewing these minutes and getting back to me immediately if there are errors or omissions. Thank you for your attendance."

## 6.2 PROJECT FINANCIAL REVIEW

### In Attendance

Frank Hamilton, Pharma Engineering Director
Edward Chase, Pharma Product Development Director
Dan Garvey, Group Engineering Manager
Gordon Taylor, Associate Scientist
Keith Carlisle, Program Manager
Jerold Workman, Pharma Financial Director

This meeting was called by Dan Garvey to review the project's 'Request for Capital Funds'(RCF).

Attendance is at the director level with only Dan and Gordon the presenters. The goal is to receive approval from the Financial Director, Jerold Workman.

Garvey had prepared the project budget for the equipment to process the drug and assemble the Oxy-Fox Inhaler. The estimate included the prototype systems and final production equipment, from the conjugation systems to the final assembly equipment and molds, and a substantial amount of engineering funding. Dan included a 15 percent capital and expense slush fund to handle equipment failures that will come up during the course of the project. The overall budget is $2.5 million in capital and $800,000 in expense. The expense budget includes $150,000 "initial manufacturing expense" (IME) to cover any operational variances in the start-up of the equipment lines. Dan's advice to the young engineers who prepare project financial estimates is "If you don't have money, you don't have friends." By this he means when things go wrong, which certainly will happen, no one will help you for free. The payback is based on estimated inhaler sales for the first two years with an 85 percent return on investment—a very good program from a financial perspective. Dan knows from past meetings with Workman that he tries to reduce program budgets by 20 percent without any consideration for the level of work or the scope involved. Workman feels that reducing funds is his prerogative and this is what he gets paid to do. Dan didn't include an additional percentage to give back to Workman.

Dan began the meeting, "You all have the RCF in front of you. If you'd like, I can go over each capital and expense item in detail. Most of these are based on engineering estimates or firm supplier quotations. For example, the prototype equipment is being purchased and modified in-house and the capital and expense required are an estimate; engineering is confident that these funds are appropriate. The final assembly equipment and production molds are from firm quotations on past projects. Any questions?"

Workman was not paying attention to what Dan was saying. He was comparing the division capital and expense plans against the money requested for this program. He answered, "Dan, I reviewed the RCF and the capital is out of line with our annual capital plan and the division is in trouble with available expense funds. You need to sharpen your pencil and reduce the capital by $500,000 and the expense by $200,000.

"I'm sure you're aware that the division prepares a capital and expense plan for the upcoming year itemizing all approved programs with the associated dollars. The Inhaler was listed with a capital of $2 million and expense of $600,000."

Dan was prepared for Workman and answered, "Jerry, good point. The capital is more than listed for this program in our annual plan. If you note the first page of the RCF, the amount allocated for this project is $2 million in the division plan. The additional $500,000 will be taken from a project that is listed in the plan but will be delayed until next year. This is allowable in our financial manual and we need these funds to do this program properly."

Jerry looked at Frank Hamilton and asked if he agreed.

Frank answered, "Yes, I agree. Dan and I went over the estimates and I concur—without these funds the project cannot be done. The project we'll delay is part of our improvement program for a utility modification at the Euro plant and is not a priority."

Workman asked Dan, "Okay. I get where the funds are coming from but I still want you to take out the $500,000."

Dan answered, "And what equipment do you suggest we don't buy? The final assembly line is a little more than the $500,000, or the final production molds . . . your choice. Without either equipment we may as well stop the program now."

Workman said, "I don't like your attitude. Your job is to design systems with the funds available. Take the money out and do your job. We could get rid of you and have the secretary do your job for that amount of capital. You are paid to do a job for 50 percent the cost that anyone else could do at 100 percent cost."

At this point Frank Hamilton had had enough. "Jerry, the program has an 85 percent financial return—the payback is less than one year. You're being unreasonable. We'll not take the capital funds out and that is final—Larry has reviewed the financials and is in agreement. And we'll have to find the additional expense dollars somewhere else. I'll take the expense dollars as an action item. Dan needs this money. Dan, continue with your presentation."

"That's all I have. If there's no more discussion, Gordon will cover the initial estimate of the inhaler cost. Gordon, you're on."

Gordon took the floor—a little apprehensive because his boss was in the meeting and he didn't know how Ed would respond. Ed had been too busy to review the cost estimate before the meeting, though Gordon had asked him to do so.

"If you look on the third page of the appendix the inhaler components are broken down individually. I have a flash drive listing each component in the inhaler and the cost."

Gordon pulled up his file on the computer and began.

Meanwhile, Ed Chase was paging through his copy of the RCF and found the appendix. He was not happy.

As Gordon was beginning his presentation, Ed jumped up, ran around the room, and closed the cover on Gordon's computer. The screen went blank and Ed yelled at Gordon.

"How many times have I told you that cost information is never to be presented without my review? Why can't you follow my direction?"

Gordon answered, "I set up a meeting but you were too busy. We can meet later today if you want."

Ed Chase sat down. He would deal with Gordon later.

Frank Hamilton looked around the room and asked the attendees, "The RCF is comprehensive of all capital and expense dollars we need to complete this program. If there's no more discussion, I would like the RCF to be approved by your group now."

Jerry Workman, upset that Frank had openly disagreed with him, said, "I'll go along. Give me a day to run sensitivity studies on the return on investment versus market volumes and then you'll have my approval."

Ed Chase, still obviously upset, just nodded. He wasn't required as an approver anyway.

## 6.3   PROGRESS MEETING: WHO TAKES CREDIT FOR WHAT?

### In Attendance

Larry Fletcher, Pharma VP Operations
Keith Carlisle, Program Manager

"Larry, we're making excellent progress. Garvey has the cocktail formulation and R&D has identified a carrier that will work in the inhaler. Garvey is building a prototype conjugation system and he's ordered equipment to dispense the cocktail into the inhaler container. There're a few technical issues that have to be resolved . . . but normal for this type program.

"Gordon Taylor's group has begun the design of the inhaler and is coordinating with Garvey. Gordon's group is reverse engineering devices that use the same technology as we need for the nasal spray and it shouldn't be a big deal. What I don't understand is why the Financial VP didn't go for the license on the Katlin inhaler. This would have been perfect—we would have a device in hand now."

Larry answered, "Workman's financial plan assumed a much lower licensing agreement and the annual cost to us to use the Katlin inhaler didn't fit. If we could somehow void the foxepin patent, we wouldn't even have to deal with Katlin or their university."

Keith continued, "Garvey is concerned about controlling a precise application of the Oxy-Fox but I'm sure he can work it out. Medical Affairs has reservations with the Katlin study but this is being worked out as well. All in all, things are moving along except for timing.

"Garvey is determined that his group needs sixteen weeks to do the prototypes for both the cocktail development and the process evaluations followed by at least a year to receive the final inhaler assembly line. I've told him to get the prototype done in ten. You know Dan—he does what he wants and doesn't cooperate—he's even started the usual friction with the materials group by insisting that his group must control the molding suppliers."

Larry hesitated for a moment in thought and said, "I've a meeting with both Frank and Jennifer tomorrow on another matter but will use the time to get the mold supplier responsibility resolved. I've about had it with Garvey's insistence on controlling the molding suppliers —Jennifer says we're leaving a considerable amount of money on the table due to Garvey's persistence that only his group can manage this. Jennifer's confident her group could get a 20 percent reduction in our molding budget if under her direct control. Any other issues?"

*Keith is confident that Garvey will work out a plan to start the final assembly line at an off-site supplier—even though he was told to keep the plan to himself,* he continues, "I've come up with a plan to reduce the time to get the final assembly in place for market launch. It's complicated but involves farming the final assembly out to a subcontractor. I'm sure that we'll pull this off and cut a minimum of two months out of Garvey's overall project schedule."

Larry smiled, "That's why you're the program manager. I expect you to identify shortcuts such as this. Good job. Go along with Garvey's prototype timing for the moment."

Larry hesitated again, considering whether he should compliment Garvey or not, finally saying, "Garvey's a good engineer and has played a major role in building this company . . . but Frank has to get him under control. Garvey is tight with Tom Watson's research group and I've tried to move him from a managing role . . . both Tom Watson and Frank support him in the group manager position and I can't buck Watson yet."

Keith continued, "Another issue is marketing. Patty is working with an outside design house in designing a device more suitable to female users and will be conducting focus groups. This could cause delays. I'll try to keep it to a minimum but may need you to intervene."

## 6.4   MORNING MEETING: JUST-IN-TIME MANUFACTURING

### In Attendance

Larry Fletcher, Pharma VP Operations
Frank Hamilton, Pharma Director Engineering
Jennifer Feddler, Pharma Director Materials Management

Jennifer Feddler controls the Materials Management group, with responsibility for procurement of all supplies required for product manufacturing, and chairs the division's cost-reduction program. Her primary responsibility is product availability at each step in the Pharma pipeline and she meets daily with both Becky Seimans and Dave Stall. Jennifer supports the product development teams: in sourcing raw materials used in development and for initial market launch; and in negotiations of all supplier contracts. Jennifer was recruited by the Pharma Division from an intravenous fluids hospital supply company. She had responsibility for all supplier coordination and contract negotiations and managed the hospital supply company's externally sourced plastic production business. There is direct conflict with Engineering over control of Pharma's sourced molding production and finished-goods inventories. Jennifer doesn't comprehend that the business is based on biological-compound processing and contends that Pharma should be run in the style of a high-volume wholesale goods store. Jennifer is an advocate of "just-in-time manufacturing" with limited pipeline raw materials and no finished-goods inventory. She contends that a pharmaceutical product should be processed only when a customer order is in queue—finished goods represent a no-interest financial investment. Engineering contends that an appropriate level of finished goods can keep a product line on the market in the face of a manufacturing issue, raw material shortage, or contamination. This position is based on comparing sales and customer goodwill lost through product unavailability versus money saved through reduced inventories.

Larry began the meeting, "Jennifer, Frank . . . Good morning. I understand that the Oxy-Fox team is doing well. I'm concerned with Garvey's timing on the prototype. Frank, you need to review this with him and get his estimate down to ten weeks . . . he's asking for sixteen weeks and that doesn't fit with my market launch schedule."

Frank knew that this would come up and answered, "I've already reviewed the timing with Dan and I support his estimate. Dan is usually right on . . . it is not the first time he's developed a process from a blank sheet of paper. In my opinion it's too early to start reducing timing on a formulation we've no experience with. I recommend that the team wait a few weeks to see how the process goes."

Larry answered, "Okay for now. Frank, another issue is the relationship between your group and Materials Management. The Oxy-Fox team is only in their third meeting and Garvey has started two arguments with Jennifer's representative, Dave Stall. It has to come to a stop . . . right now."

Frank answered, "Larry, I spoke with Dan. There're two sides to this story. Molding requires a strong technical lead from our company, which I support. Jennifer's group does not have these skill sets."

Larry looked at Jennifer, "Any comment?"

"As a matter of fact I've a lot to say. First, the technical knowledge is really with the molding supplier. We don't have to provide this direction. These suppliers are competent to handle the business with minor direction. I did this function at my previous company and was rather successful, resulting in a 20 percent reduction in the overall cost of the purchased commodities. I can do it here as well."

Frank started to explain the technical role provided by Engineering but Larry, looking at his watch, stood up, "I have to meet with Siegal in fifteen minutes and need time to prepare. You two get together and work it out."

Frank Hamilton left the room while Jennifer hung back, "Larry I need some time on your schedule. The Pharma group is growing to the point that it's time to consider reorganization. I have a plan that I'd like you to review."

"I understand. Get together with my admin and set it up."

# 7

# CGMP PROCESS VALIDATION REQUIREMENTS

## 7.1  LAUNCH TEAM MEETING NUMBER 4

**In Attendance**

Keith Carlisle, Program Manager
Susan Jaffey, Quality
Dan Garvey, Engineering
Gordon Taylor, Product Development
Dave Stall, Materials Management
Jeffrey Daniels, R&D

**Not in Attendance**

Dr. Gonzales, Medical Affairs
Patty Keyser, Marketing

Keith Carlisle presented the agenda for this meeting with action items identified from team meeting number 3.
Keith asked, "Dan, have you reviewed the sixteen-week schedule?"
"Yes I have and I am sticking with my initial timing."

---

*A History of a cGMP Medical Event Investigation*, First Edition. Michael A. Brown.
© 2013 John Wiley & Sons, Inc. Published 2013 by John Wiley & Sons, Inc.

"That is not acceptable. You need to take six weeks out. You put this time in to handle any failures in the prototypes. I pay you not to have failures."

Dan responded with a tone that made Keith flinch, "For one thing, you do not pay me. If directed by my boss I'll take the time out. But I assure you that there will be failures—I've never worked on a design that worked perfectly. Modifications are always required. If the six weeks are taken out, I assure you that the time will be needed and the schedule will not be met. Think about this—isn't it better to plan than to react?...By forcing a schedule reduction you'll be putting the entire organization in a reaction mode. If the prototypes are characterized incorrectly, the actual production equipment will not produce acceptable product. I plan for a 90 percent success in design, allowing reaction time to resolve a 10 percent failure. If you would rather plan for a 10 percent success with a 90 percent reaction time, . . ."

Keith stopped Dan in midsentence, "Enough. We'll go with the sixteen weeks for the time being. Gordon, we don't have to review your timing. Management feels the time is appropriate."

Looking at the agenda for new items, Dan saw his presentation was the only agenda item and he began an explanation of the qualification requirements.

"I know the team has heard more than enough of Food and Drug Administration procedures, but I need to cover the basics needed to meet Agency requirements. As an organization our equipment and process verifications and validations are based on Good Automated Manufacturing Practice—or GAMP—recommendations.

"Equipment, utilities, and facility qualifications are required by Current Good Manufacturing Practice—or cGMP—Regulations for Finished Pharmaceuticals. Our inhaler is subject to these requirements; Medical Affairs did not determine if a premarket approval—a PMA—for devices is required or not . . . makes no difference . . . basically the same rules. The inhaler is a class 3 medical device and must comply no matter who says what to the contrary.

"We are required to have written procedures for production and process control designed to assure that the drug products have the identity, strength, quality, and purity they purport or are represented to possess. This is to demonstrate the effectiveness and reproducibility of the process.

"We need to establish control procedures to monitor the output and validate the performance of our manufacturing processes that could introduce variability in the characteristics of in-process material, our Oxy-Fox cocktail, the conjugation and dispensing-and-capping equipment, and the assembled nasal inhaler . . .as well as establish written procedures designed to prevent microbiological contamination of the cocktail, both for the production lines and for the facility. The inhaler components will be injection molded in a clean-room-type facility and assembled either on our site or at the molder. If not initially, then longer term, we will definitely assemble at the molding site as a cost reduction. Assembling at the point of molding will give us points toward our lean-manufacturing goals. At this point in the development, the cocktail will be produced under aseptic conditions. Following identification of the process parameters for production of the cocktail a Validation Master Plan—known to you as the VMP—will be written.

"The VMP serves as both a guide for our organization and a review document for the Agency. Since the document may be used by organizations unfamiliar with the ongoing activities required to qualify the equipment, utilities, and facilities, the VMP must be fairly comprehensive. The plan describes the critical utilities and equipment and the qualification methodology, with preapproved protocols for each area and piece of equipment. Specific acceptance criteria for all critical utility systems and equipment will be briefly described. Additional programs to establish ongoing control, such as training, calibration, and preventive maintenance, will be part of the master plan. The VMP will undergo a technical and compliance review by the Quality group and must be approved prior to the preparation of validation protocols. For a small project no master plan is required; our Equipment Change Control System, which is as you know part of our CAPA plan, would suffice . . .which is not the case for this program."

Dan paused, looking for any reactions; he noted that Keith was confused.

"Keith, what do you not understand?"

"I do not know what our 'CAPA' plan even is."

Dan continued, "The 'CAPA' stands for 'Corrective and Preventive Actions' and the CAPA plan is a major part of our Quality System. Okay? Back to the VMP. Specifically, the VMP will include validation documents describing the total in-house process, including the conjugation process, dispensing-and-capping equipment, inhaler assembly line, and the facility's environmental control. Supplier qualifications for the molded components will be noted in the VMP but performed by the molding company engineers under the guidance of our molding engineering group.

"Each subsequent validation for the individual qualifications will assure that the equipment and process are designed or selected so that the product specifications are consistently achieved. These documents will be prepared by Engineering with input from Sue and the operations manager at the corporate plant. Keith, please note in the minutes that we must get a plant representative on the team."

Keith responded, "Noted."

Dan continued, "In Agency terms the validations will be in the form of a Prospective Validation that in our case must be completed prior to introduction of the inhaler or when there is a change in the manufacturing process—which may affect the inhaler's characteristics, such as uniformity and identity.

"Key elements of the Prospective Validation are the Design Qualification, abbreviated as DQ; Installation Qualification, abbreviated as IQ; Operational Qualification, abbreviated as OQ; and Process Performance Qualification, abbreviated as PQ. The results will be used as objective evidence to support the overall Verification and Validation documents submitted with the New Drug Application—the NDA. Each qualification document will consist of a preapproved protocol prior to commencement of any qualification activities. This is an important point: what we plan on doing must be written and approved prior to actually doing the qualifications.

"The aim of the Design Qualification is to establish and document that the design of the equipment is suitable for the plant's intended use. The document includes a package consisting of User Requirements, Design Specifications, Functional Specifications, and Automated Controller Specifications.

"The IQ must be a preapproved protocol with specific objective evidence that will be gathered to assure that the facilities, utilities, and process equipment have been installed and meet the established design criteria. Some of the equipment will be purchased and used without modification, and the IQ for this equipment will be based on the supplier specifications.

"A separate IQ will be prepared for each individual critical utility system or piece of equipment and referenced in the Master Plan. This includes the room that the conjugation process is in, the conjugation equipment itself, the container dispensing-and-capping system, the inhaler assembly equipment, and the utilities that support the operation of each. During the IQ phase a complete review of the utility system or equipment prior to start-up will be performed. The preapproved protocol will provide a systematic method to check the system's static attributes prior to normal operation. A detailed description of the system will be required. The description will include an explanation of what the system is intended to do and all major components of the system. The system will be reviewed postinstallation to verify that the system installed is the same as that which was specified."

Dan paused, but only to catch his breath. "The locations of documents such as specifications, engineering drawings, manuals, and data sheets will be referenced. In the past these documents have been kept in the engineering library. We now have a locked room controlled by the Quality group to house all our project documentation. The system will be examined for proper connection and installation of supporting utilities. Model and serial numbers of equipment will be recorded. Any deviation from the preapproved protocol requiring a change in the qualification package will be investigated and the appropriate course of action determined, including justification to make the change, the implemented correction, and the requalification of the system. All change data must be included in the qualification package as an addendum. Each change requires review and approval by the Quality organization.

"A simple explanation is the IQ tests equipment for fit and function, so to speak—that is, is the equipment installed according to design specifications, does it run, and when it runs does it do what it is supposed to do? An example is a test on a valve: is it the specified valve, is the power requirement as specified, and does the valve actuate as required in the design?"

Dan stopped and asked if there where any questions; he was met with only blank stares, so he continued, "Information obtained from the IQ is used to establish written procedures covering equipment operational specifications, calibration, maintenance procedures, repair parts lists, monitoring, and control. The objective is to assure that all repairs can be performed in such a way that will not affect the characteristics of product processed after the repair. In addition, special postrepair cleaning and calibration requirements will be developed to prevent inadvertent manufacture of nonconforming product. Any questions up to this point?"

Keith Carlisle answered, "Not really. Please continue."

"Following the completion, review, and approval of the IQ, the Operational Qualification can begin. Basically, the OQ establishes and documents by objective evidence that the facilities, utilities, and process equipment perform as intended throughout representative or anticipated operating ranges. The OQ is a preapproved protocol as well, and any deviation encountered during performance of the OQ requires the identical investigation and approval as the IQ.

"The OQ entails rigorous testing to demonstrate the effectiveness and reproducibility of the process. In entering the OQ phase of validation, the process parameters such as operating temperature, dispensing accuracy and capping pressure, and Oxy-Fox concentrations determined in our prototype systems will be tested and challenged to assure that the process is capable of repeatably meeting these requirements. It's important that the equipment qualification simulate actual production conditions, including those which are worst-case situations. The identification of worst-case conditions will be done on the prototype process equipment and one of the outcomes of the process failure mode effects analysis—the FMEA. Raw material lots used in the OQ may also be used in the Process Qualification, which I will address next. A minimum of three runs will be required as part of this phase.

"The OQ will describe operational tests and measurements of the parameters critical for the proper operation of the system. Test objectives, methodologies, and acceptance criteria will be defined and preapproved as in the IQ. Calibration of the system will be documented during the OQ. Controls will be manipulated to test the ability of the system to provide adequate control over process parameters. All indicators, recorders, alarms, and interlocks will be monitored as well. Peak load challenges—such as running the assembly equipment at a higher throughput than encountered in regular production—will be defined and incorporated into the testing strategy, to challenge the system's capability. It is important to point out that all instrumentation used to control or monitor the process parameters must be calibrated prior to the beginning of the OQ and recalibrated following the completion of testing. This includes all sensors and gauges.

"Now, a critical phase of the qualifications, and one that is a bone of contention between our group and the Information Technology ... IT ... group, is the software validation. The equipment and process will be controlled by microprocessors called programmable controllers. The programming for these units really follows the operation of the equipment. In the opinion of the Engineering group, if the equipment meets all qualification challenges, the programmable controller is functional as well and the completion of the OQ also qualifies the programmable controllers. This meets the validation requirement. As I said, there is disagreement between our group and the IT organization. The IT group is accustomed to writing their own code, which does require separate validation, and do not understand that programmable controller code is written by the supplier and validated by the supplier.

"As part of the OQ, operating procedures for each piece of equipment will be defined and described with sufficient specificity so that operations personnel understand what is required to sustain the equipment and process at designed operating conditions. All operators will be certified by examination that they understand the operation of the equipment prior to assignment to the production line. In a nutshell so to speak, the OQ determines that the operation of the equipment meets design specifications. As an example, the filling equipment statistically meets the requirement that it will fill the appropriate volume into the container as specified. Any questions at this time?"

Dan looked over the group and just saw all heads nodding. Dan continued, "The Process Qualification, or PQ, follows the completion and approval of the OQ. The PQ is a preapproved protocol that has identical discrepancy documentation requirements as the IQ and OQ. The PQ will be used to test the performance or process parameters of each system that could affect the product. The PQ will integrate procedures, personnel, materials, equipment, and process. A PQ establishes the test objectives, methodologies, and acceptance criteria defined and approved prior to execution. A sufficient number of replicate studies will be performed to determine the ability of each of our processes and assembly equipment to achieve reproducible results. This phase of the validation requires the production of actual product. Testing may include analysis for chemical, physical, and microbiological constituents. The protocols will incorporate peak load challenges to the intended operating range of the system or process as defined in the OQ."

Keith Carlisle interrupted asking, "Could we use a retrospective Performance Qualification to reduce the overall qualification time?"

Dan answered, "Retrospective Qualification is not an appropriate path for a new product introduction and the Agency frowns on this type plan. If a product has been on the market without sufficient pre-market process validation a retrospective qualification may be appropriate. In these cases, it may be possible to validate, in some measure, the adequacy of the process by examination of accumulated test data on the product and records of the manufacturing procedures used.

"Retrospective validation can also be useful to augment initial pre-market prospective validation for new products or changed processes. In such cases, the prospective validation should have been sufficient to warrant the product launch. As additional data is gathered on production lots, such data can be used to build confidence in the adequacy of the process. Conversely, such data may indicate a declining confidence in the process and a commensurate need for corrective changes. In addition, we may be required to requalify each line every two years—in this case I believe a Retrospective Validation package may suffice—though there is some disagreement from the Quality group on this point . . . Keith does this answer your question?"

Keith just said, "Yeah."

Dan continued, "I'm almost done . . . just need to cover a few more points in the Process Qualification phase of our validation program.

"The PQ involves the evaluation, sampling, and testing of the assembled Oxy-Fox Inhalers at critical process steps to establish objective evidence that provides a high degree of assurance that the Oxy-Fox conjugation process, container dispensing and capping, and final inhaler assembly will consistently produce a product meeting the predetermined specifications and quality attributes expected by our customers.

"As mentioned previously a sampling plan will be determined by the Statistical Methods group based on an evaluation of the process to be validated, with designated critical process steps receiving the most attention. All finished inhaler lots will be placed on stability."

Keith interrupted, "The expiration date on the inhalers will be 12 months to start. Does this mean that we cannot enter the market until this is complete?"

Dan answered, "Keith, the inhalers produced on the prototype equipment can be used in the stability studies and proceed concurrently with the design and installation of the final production equipment. This is another reason why the final production equipment must follow the same validation plan as the prototype conjugation processes, dispensing-and-capping processes, and inhaler assembly equipment. Though there is a risk in doing this: If a deviation is required in the qualifications of the final production equipment that affects the functionality of the inhalers, the stabilities will be invalid. This is why my group must spend the sixteen weeks that I've asked for to assure the process parameters are finalized so that there will be no changes in the future production equipment. Any more questions?

"Okay, I'll continue. In the PQ, it is important that challenge conditions simulate those that will be encountered during actual production, including worst-case conditions. As in the OQ, these conditions will be identified from the prototype equipment and in the FMEA documentation. The challenges should be repeated enough times to assure that the results are meaningful and consistent and include a minimum of three production lots.

"Again, acceptance criteria will be established by the Statistical Methods group with set levels of how many finished parts are rejected and any nonconforming product found in subsequent inspection . . . a random sample of finished inhalers will be used to test actual functionality; that is, do these inhalers deliver the specified dosage of compound upon actuation and does the inhaler contain the total amount of compound as specified in the product labeling? In our case each inhaler is designed to deliver 34 applications. There will be a minimum of three qualification runs and possibly more. Depending on the confidence level set, we'll be allowed a certain number of failures. Normally the confidence level is 95 percent."

Sue raised her hand, "95 percent seems rather low. I think we should go for 99 percent."

Dan answered, "Sue, 99 percent requires a huge sample size. Probably more than we could even sell the first year. Many of the inhalers would have exceeded the expiration date by then. I would like you to set up a presentation from the Stat group to answer this type of question."

Sue continued, "I'll do that but I need Marcia's opinion on this before I'm willing to accept a 95 percent confidence level."

Dan answered, "Sue, no problem. Keith, can you set this meeting up?"

Keith Carlisle responded, "Sue, you do this."

Dan was getting tired, "If there are no more questions I'm almost done. That in a nutshell is the extent of the validation requirements.

"Now as part of the Verification and Validation plans it will be necessary to demonstrate that both the prototype and final processes do not adversely affect the finished product. The objective evidence from our PQ includes product testing using actual inhalers manufactured from the same type of production equipment, methods, and procedures that will be used for routine production. This supports our plan to provide inhalers for the clinical studies from the prototype systems. If the prototype systems are not shown to be equivalent to the final production process, the product may not be representative of product used for market launch and the clinical studies will be invalid.

"This would be a worst-case scenario for all our careers."

Keith gave Dan a thumbs-up, "Good point."

Dan continued, "After actual production units have successfully passed the PQ, a formal technical review at the Pharma Director level will be conducted including a comparison of the approved product specifications and the actual qualified product, determination of the validity of test methods used to determine compliance with the approved specifications, and determination of the adequacy of the specification-change control program. This is the final Tollgate prior to the initiation of our clinical studies.

"Note my last point. The specification-change control program at our Argentina plant has been an Agency audit issue in their change control documentation system. We'll have to do a better job when the inhaler is transferred."

Dan stopped again to catch his breath, "I didn't plan this presentation to go into as much detail as it has and I appreciate your patience in not interrupting too much. I think that the necessity of validating the prototype systems for the inhalers used in the clinical studies has been answered. Any questions?"

Keith wanted to get this meeting over; in his opinion Dan was starting to just ramble, "Dan . . . thanks. I know you put a lot of time into gathering this information and I understand the importance of validating or qualifying the prototypes . . . I don't understand the use of the two terms interchangeably. But I still want a ten-week program. You really need to gather your soldiers and take another look . . . and assume that the qualifications can go in parallel with the clinical studies. Sue, do you disagree?"

"Actually, I do disagree. You have been told but obviously are having trouble accepting that the Agency requires the clinical lots to be produced on validated equipment. The Quality group will insist that the qualifications be completed prior to release of the Oxy-Fox Inhalers to Medical Affairs and the qualification documents fully reviewed and approved."

Keith sat silent for a moment, knowing if the prototype processes had to be fully qualified with approved documentation prior to release of the inhalers to clinicals, the project would not meet Larry's schedule, but he kept silent—he had to at this point. "That's more than enough for today. We need to take all that Dan presented and put it in a schedule form."

Keith continued, "I'll publish meeting notes with agenda items for the next meeting. Sue, at the next meeting I would like you to update us on what a FMEA involves. Thanks."


\*\*\*


Dan went back to his office to find a note from Debra Wonzer, one of the other program managers who deal with product launches. Debbie wanted him to call her as soon as possible. An illegible scribble on the note looked something like needing help with a new computer. Dan called her and she asked if he could come to her office and help her load some software.

Dan knew that Debra Wonzer was new to the program management team and hasn't had responsibility yet for any market launch programs. Larry assigns special projects to Debra; he needs to be convinced she'll follow his instructions without reservation before giving her any real responsibility.

Debbie's office is on the second floor of one of the older buildings on the other side of the Kinnen campus. Dan had things to do and didn't want to take the time to go over to her building, do whatever she needed with the computer, and come back. Parking was a problem and the campus shuttle had become totally unreliable. Dan went anyway. Debbie had been nice to Dan the few times they'd met. She had a friendly smile and what guy can reject a request from a girl who smiles.

Dan drove to her building, finding a parking spot in front . . . unusual but it was slightly after regular hours and he figured many of the people had gone home. The elevator was slow . . . one of those old hydraulic types that take forever . . . so he walked to the second floor. No big deal. The offices in the newer buildings have glass walls with blinds that can be positioned for privacy. Blinds are normally left open. The offices in this building had plaster walls with wooden doors. Dan couldn't see into any office and was not certain which office was Debbie's. The name tags beside the doors gave it away in a flash. Her office was at the end of the corridor . . . pretty much isolated.

Dan knocked on the door to get Debbie's attention. "Dan, come in. I'm so glad that you could come over and help."

Dan went into her office, leaving the door open. There was no need for privacy.

Dan asked, "What's up with the computer?"

"I have this CD that won't load. I've tried a hundred times and all it does is give me an error message."

"Let me try." Dan loaded the disk. "Okay . . . same thing."

Dan shut the computer off through the operating system and restarted it. It took slightly longer than expected to open. He tried to load the CD—with the same results and same error message. Dan looked at the properties of the CD and noted that it was in the 250-megabyte range, then looked at the available memory on the hard drive, which was about 200 megabytes. "Debbie you don't have enough memory to load this program. If you really need it, you need another computer . . . or take some of the stuff you no longer use off your hard drive. You need to delete about a gig of your programs."

Debbie answered, "I need to get this program running tonight. No time to get another computer. Can you help me drop stuff off the drive?"

Dan thought that this would take more time than he had but answered, "Sure."

Dan sat with Debbie for an hour while she decided what programs could be deleted and what she had to keep. Dan was patient. Debbie was a nice lady. Dan did not mind spending time to help her.

When they had made room on the drive, Dan loaded the new program and was making sure it ran OK.

Debbie's office was rectangular, not the square design in the newer buildings; her desk was on the opposite wall from the door. While Dan sat at the computer, with his back to the door, Debbie shut the door, rejoining Dan at the computer. She sat close to Dan. She sat closer than she had while they were working on the drive—so close that Dan became a little uncomfortable. Dan's left hand was resting on her desk while he maneuvered the computer mouse with his right.

Debbie placed her right hand next to his left, very slowly running her little finger over the top of his hand, and seductively asked, "You've been so nice, is there anything I can do for you . . . anything?"

Completely startled, Dan jumped up and said, "Okay, we're done . . . computer works fine . . . got to go," and left Debbie as fast as he could, almost forgetting to open the closed door.

Walking back to his car Dan thought, almost out loud, "*Good-looking women like Debbie never really come on to me. She came on to me! Has to be a setup . . .*"

Dan walked even faster.

Dan left the building without stopping and stood by his car for a moment, letting the evening air dry the perspiration on his face. He then went straight home.

# 8

# FAILURE MODE EFFECTS ANALYSIS

## 8.1   LAUNCH TEAM MEETING NUMBER 5

### In Attendance

Keith Carlisle, Program Manager
Susan Jaffey, Quality
Dan Garvey, Engineering
Gordon Taylor, Product Development
Dave Stall, Materials Management
Jeffrey Daniels, R&D
Lynn Diehl, Plant Operations

### Not in Attendance

Dr. Gonzales, Medical Affairs
Patty Keyser, Marketing

Keith presented the agenda for this meeting; no action items identified from team meeting number 4 were listed, to Garvey's surprise. The agenda included an introduction of the plant representative and Sue's review of failure mode effects analysis (FMEA) requirements.

*A History of a cGMP Medical Event Investigation*, First Edition. Michael A. Brown.
© 2013 John Wiley & Sons, Inc. Published 2013 by John Wiley & Sons, Inc.

Keith Carlisle opened the meeting, "I want to introduce Lynn Diehl from site plant operations. Lynn reports to Bruce Garlin, Site Plant Manager. Lynn, what will be your responsibility on this team?"

"The prototype process and assembly equipment as well as the final production equipment will be run by my department. I'll provide the operating personnel for the process parameter development and subsequent qualifications."

"Glad to have you on the team." Keith looked at Sue and said, "Sue, let's get started."

"Okay. Failure mode effects analysis is a big area and I'm not sure how detailed we want to get into this at this time."

Dan Garvey responded, "We need more than just an overview as this is required by our Six Sigma program and I'm not sure all the team members are familiar with the procedure."

Sue continued, "I've a presentation prepared but it will take the full meeting time to go over. Keith, is it okay with you?"

"Be my guest, Sue; go ahead."

"Okay. To start, as you know FMEA stands for 'failure mode effects analysis' and encompasses more than one type of analysis.

"An FMEA provides the design engineer, manufacturing engineer, and others a systematic technique to analyze a system, subsystem, or item, for all potential or possible failure modes. This method then assigns a probability that the failure mode will actually occur and what the effect of this failure is on the rest of the system. The criticality portion of this method allows one to place a value or rating of how seriously the failure affects the entire system.

"The following administrative information is required to prepare the FMEA: a description of the device, design responsibility, team members, the person responsible for FMEA preparation, and the date the FMEA was prepared and its revision level. An FMEA is an evolutionary process—as new information is determined the FMEA may require review and modification.

"Technically, the FMEA addresses the subsystem or component under detailed analysis: the component function, the potential failure mode, the potential effect of failure, the potential cause of failure, and identification of the current modes of control in place to prevent the cause from occurring. If the modes of control are not adequate, the team needs to identify potential modes of control that can be implemented to prevent the cause from occurring. From this technical evaluation, a risk assessment is performed and a risk priority number—an RPN—is determined. The RPN is a dimensionless number so there is no real meaning to a value of, say, 600 versus 450 except the difference in magnitude. It is up to the FMEA team to interpret what RPN magnitude represents a risk.

"The next major step is to weigh the risks associated with the current component, effect, and cause, with the modes of control that are currently in place or could be put into place. The severity value of the effect of the failure on the rest of the system is then assigned. The severity value is indexed from 1 to 10. A value of 1 means the user will be unlikely to notice the failure and a 10 means that the safety of the user is in jeopardy.

"A weighting is then assigned to the probability the failure mode will occur. Values for this probability index generally range from 1 to 10, with 1 being virtually no chance and 10 being near certainty of occurrence.

"A measure of the effectiveness of the modes of control currently in place is then determined that identifies or mitigates the potential weakness or failure prior to release to production or to the market. This index also ranges from 1 to 10. A value of 1 means it will certainly be caught, whereas a value of 10 indicates the design weakness would most certainly make it to final production or be released to the market as finished goods without detection—this would be nonconforming product and in our case could cause an adverse patient response.

"The risk priority number is the product of the indices from the three previous assessments." Sue stopped at this point and wrote on the whiteboard:

$$RPN = Probability \times Severity \times Mode\ of\ Control$$

Sue then continued, "The next activities are based upon what items have the highest RPN and/or those with the major risk issues. This is somewhat arbitrary and requires technical knowledge of what the actual consequences will be in case of failure. A high RPN does not necessarily represent an issue that must be addressed. It is strictly a technical and team decision.

"When the risk is determined and the team decides that a system modification or corrective action must be incorporated into the design or the manufacturing operation to reduce the risk, the corrective action must be evaluated and a new mode of control identified along with the person or group responsible for implementing the corrective action. The final step is a calculation of the revised RPN based on implementation of the corrective action.

"In summary, the FMEA provides a disciplined approach for the project team to evaluate designs to ensure that all the possible failure modes have been taken into consideration. Any questions at this time?"

There were no questions. Keith Carlisle asked the team, "Do you think an example would be appropriate for clarification?"

Dan answered that he had a simple example and would wait until Sue was finished.

Sue continued, "I want to spend some more time on the risk assessment as it is the most important part of the analysis.

"Risk assessment is the combination of the severity of failure and the probability of a failure. The severity of failure is the effect of the failure on the system, operators, or system performance. The probability of failure is the likelihood of the failure occurring. The likelihood of detection is not considered in the risk assessment.

"The risk assessment is an alternate way of looking at the RPN and is more appropriate to a manufactured medical product than the measures of severity, probability, and mode of control I just discussed. The end product may be the same but the approach is a little different. To repeat, the risk assessment is based on the severity of failure and the probability of failure. If either of these two indices is determined by the team to be unacceptable, that is, if a patient injury would surely occur, no mode of control may be adequate and a redesign of the product may be required.

"Severity of failure, in our case, has five categories. The highest concern is 'catastrophic,' meaning the failure may cause patient death or severe damage to the inhaler and is indexed as a 10. Catastrophic is followed by 'critical,' meaning a failure may cause severe injury or major damage to the inhaler, and is indexed from 7 to 9. Next is 'marginal,' meaning a failure may cause minor injury to the patient or degradation of the inhaler with an index of either 5 or 6. And then 'minor,' meaning failure will not cause patient injury or inhaler damage although it may result in a failure in manufacturing and unscheduled maintenance but the failure is discovered prior to release. Minor is indexed between a 2 and 4. Also, a failure that has no impact on performance is indexed as 1. Note there is no zero index.

"The probability of failure is also ranked. A common ranking of failure probabilities is a 10 for a high likelihood of occurrence—statistically meaning the failure could occur in 1 out of 3 products. A ranking between 7 and 9 indicates a probable occurrence in 1 out of 20 products. A ranking between 4 and 6 indicates an occasional occurrence such that a failure could occur in 1 in 2000 products. And a ranking between 2 and 3 indicates a remote probability that a failure will occur in 1 in 150,000 products. The lowest probability of failure is ranked as

1 for a failure that is highly unlikely—or 1 failure in 1.5 million products. This is not a perfect system as you can see. A failure of 1 in 150,000, though remote, still could result in a serious patient injury and we classify it as an acceptable risk.

"Detection is similarly categorized: from being absolutely certain that the mode of control will not detect the failure, leading to an index of 10, to being almost certain the mode of control will detect the failure, giving an index of 1. A mode of control that has a remote chance of detecting the potential failure has an index of either 8 or 9. A detection method that has a low chance the mode of control will detect a potential failure has an index from 6 to 8. A detection method that has a moderate chance the mode of control will detect a failure is indexed as either a 4 or 5. A detection method that has a high chance the mode of control will detect failure is indexed either as a 2 or 3, and a detection method that is almost certain to detect the failure is ranked as a 1."

Keith Carlisle asked, "Sue, I don't follow the multiple rankings. I mean, once a severity or probability category is determined how do you make the decision whether an index is ranked, say, 6, 7, or 8?"

Sue answered, "Strictly a technical decision. The team doing the FMEA must have significant technical knowledge of the product, process, and consequences of failure to establish these ratings. Obviously, the medical representative will have to show up at the meeting when we do these assessments.

Sue continueed, "When a failure is identified and a mode of control is determined to be effective, a new RPN is determined based upon a new mode of control that lowers the RPN value to an acceptable limit. Depending upon the RPN value and its interpretation in improving the detection or the occurrence of the failure, the new mode of control will have to be implemented prior to product launch and qualified.

"There can be four different FMEA analyses used in the program at subsequent stages of the process and product design phases. These include the design, process, system, and functional FMEA.

"I know I'm taking a lot of time but I want to briefly explain each of these FMEAs.

"Design FMEAs are performed on the process at the design level. The purpose is to analyze how failure modes affect the product and system, and to minimize consequences of the failure. The FMEAs are completed before the process is released to the plant operations. All anticipated design deficiencies will have been detected and corrected by the end of this process.

"Process FMEAs are performed on the manufacturing processes. They're conducted through the quality planning phase as an aid during production. The possible failure modes in the manufacturing process, limitations in equipment, tooling, gauges, operator training, or potential sources of error are highlighted and corrective action taken.

"System FMEAs comprise part-level FMEAs. By part level, I mean each subcomponent of the process or product is evaluated. As an example, take the nitrogen valve on the conjugation system. A part-level FMEA evaluates what could go wrong with the valve, the effect on the product, and the mode of control in place to prevent its occurrence. All of the part-level FMEAs tie together to form the system. As a FMEA goes lower into the system, into more detail, more failure modes will be considered. A system FMEA need only go down to the appropriate level of detail."

Dan interrupted, "Sue, the system FMEA goes into too much detail and takes an inordinate amount of time and effort from both our groups, and the information obtained is usually not pertinent and more than often identifies what I call 'ghost issues' that can't be corrected anyway."

"Dan, the system FMEA is your responsibility and no longer required by our Six Sigma program, as you know. If your group doesn't see any benefit in it . . . don't do it.

"Functional FMEAs are also known as black-box FMEAs. This type of FMEA focuses on the performance of the intended finished device, which is our Oxy-Fox Inhaler, rather than on the specific characteristics of the individual parts. As an example, if a product is in the early design stages, a black-box analysis would focus on the function of the device rather than on the exact specifications. This analysis would focus on the inhaler's ability to meter the cocktail on patient administration and not the color, dimensions, or the components of the device.

"And finally, all of these FMEAs can be applied to software systems as well. I'm sure each of the process prototypes and final production equipment will be operated by some type of software . . . That's it for now. As we get further into the program we'll go over the FMEA requirements in even more detail."

Keith asked Dan if this was a good time to cover the example he had in mind.

Dan began, "I've put up a PowerPoint of a design FMEA for a coffee cup." (See Fig. 8.1.) "This is simple and should clarify the main points in Sue's presentation.

"The system or product is a coffee cup. The function includes the CTQs noted as critical customer requirements, or in Agency terms,

| System | Function | Potential Failure Mode | Potential Effect of Failure | Potential Cause of Failure | Current Mode of Control | Risk Assessment | | | | Additional Mode of Control and Action Taken | Revised Risk Assessment | | | |
|---|---|---|---|---|---|---|---|---|---|---|---|---|---|---|
| | | | | | | S | P | D | RPN | | S | P | D | RPN |
| Coffee cup | A device to hold 10 ounces coffee maintaining a temperature of 72 to 90°F for approximately one hour with no spillage while driving a car | Leakage | Burns to user | Fracture of device | Design | 8 | 1 | 1 | 8 | Redesign to include a hinged lid Responsibility: Product Development | 8 | 5 | 5 | 200 |
| | | | | Lid not on tight | Instructions for use | 8 | 7 | 10 | 560 | | | | | |
| | | | | | Design of easy on – somewhat hard off | 8 | 7 | 10 | 560 | | | | | |
| | | Spillage | Burns to user | Failure to install top | None | 8 | 5 | 10 | 400 | Hinged lid | 8 | 1 | 3 | 24 |
| | | | | Dripping | | | | | | | | | | |
| | | Temp control | Cold coffee | Time | Material selection | 2 | 5 | 5 | 50 | Hinged lid | 2 | 2 | 2 | 8 |
| | | | | Lid left off | | | | | | | | | | |

**Figure 8.1** Design FMEA for a coffee cup.

design inputs, and are noted as 'A device to hold 10 ounces coffee maintaining a temperature of 72 to 90°F for approximately one hour with no spillage while driving a car.'

"One potential failure mode is leakage, with the potential effects of failure noted as burns to the user and cause of failure noted as fracture of device or lid not on tight. The current modes of control identified are design, instructions to user, and lid design: easy on, somewhat hard off. In the risk assessment, 'S' identifies the severity of the failure, 'P' the probability the failure could actually occur, 'D' the assessment of whether the failure can be detected by the mode of control prior to product release or identified by the customer, and the computed RPN. Each factor in the risk assessment is somewhat arbitrary, as Sue presented, and depends on the technical experience of the team members.

"In this example the coffee cup could easily result in spillage in a moving car if the lid is not securely in place. The first potential cause of failure is fracture of the coffee cup: it has a severity of 8 as a result of burns to the user; the probability of the material fracturing is low so the index is 1; and the mode of control is design of materials and is rated as a 1 as well. The lid not being on tight and spillage occurring has a severity index of 8 as this may result in burns to the user; it is assigned a probability of occurrence of 7 because there are no modes of control in place that could eliminate this issue; it is assigned a level of detection of 10 since the spillage will only be detected after the spill has occurred. The associated RPN is the product of these three factors—in this case 560. The high RPN shows that the mode of control, instructions for use, will not be effective in eliminating the spillage. A new mode of control requires a redesign incorporating a hinged lid, for which the severity remains at 8, the probability of occurrence is reduced to a 5, and detection is also reduced to a 5, thus reducing the RPN to 200. If the hinged lid is not installed on the coffee cup in production, it can be easily detected but does not guarantee that the user will actually close the lid. Not a perfect control but a significant improvement. The same logic follows for the additional failure modes listed in the overhead. Note the last potential failure mode, temperature control. The potential effect of failure is cold coffee, with a severity of 2; this rating reflects user dissatisfaction but will not result in injury. The potential cause of failure is time or lid left off, with a rating of 5—time and use are out of our control. The current mode of control is material selection. The material would be specified based on thermal conductance for, say, a minimum of one hour, and the mode of control is design. The risk assessment ranking is 50. The addition of a hinged lid reduces the risk assessment to an 8.

Dan paused to register any reaction from the team, and then continued, "One point that's important—and has put me in front of the management committee before—is that FMEA incorporates the failure modes the team identifies. In my experience these failures do not occur in practice because they were identified and modes of control were put in place. There will be failures that the team fails to identify and the management team will get all over our case if these occur. This is my biggest issue with the FMEA process. If we do a thorough job in identifying failure modes, 90 percent of potential events will be eliminated—but we'll be accountable for those not identified that may happen. Any questions?"

Keith answered, "I think we get the idea. This could be rather involved for the inhaler. The team should begin thinking how the risk analysis should be implemented. Sue, is your group responsible for managing this function?"

"Probably, and obviously we will need the input from the entire team. And as I said, please make sure Medical attends."

Keith answered, "No problem."

"That's about all the time we have; again we're a few minutes over schedule and have to get out of the room. I'll schedule a meeting for early next week. I want an overall schedule for the entire program—including prototype design, process parameter development, qualification, component design availability and when the molding supplier will complete the final process design, when the preproduction molds will be available and when the molded components will be in-house, when the pressurization system will be delivered and operational, when finished inhaler assemblies will be available for clinical studies, when the process and assembly production equipment will be sent out to bid including the evaluation of the bids and award, when the production equipment will be received in our plant, the time for all the qualifications, and when finished product will be available for market release. Do not address transfers to either Argentina or our Euro plants at this point."

Dave Stall asked if Keith had any information from Marketing on the quantity of inhalers for the initial market launch. Keith didn't know and would make sure Patty would be at the next meeting. Dan also asked if Medical Affairs had given any information on the clinical studies and how many inhalers would be needed. Keith answered he would also contact Dr. Gonzales.

\*\*\*

It was still early in the day and Dan and Gordon were scheduled to attend the first inhaler design meeting but Dan needed to check his messages first. Back in his office there was a note from Frank that he wanted to talk to Dan as soon as possible. Dan asked Frank's admin, "Who have I upset now? Is Frank available?"

Frank was in his office and Dan knocked on the door.

"Dan, I need your advice on a meeting I just came from. You need to keep this strictly confidential."

Dan took a chair at the conference table. Frank continued, "Dan, as you know, corporate hired a management consultant to review Larry's organization. Most of his time is spent resolving product availability issues and plant personnel problems, not to mention the time he has to spend meeting with his nineteen direct reports. He's overwhelmed. Corporate believes he should be focused on the division's new product goals and sales.

"To this end, the consultants interviewed each of us—all nineteen reports—for suggestions. From the meeting it's obvious the consultants spent much of their time with Jennifer. I know that she had privately presented her version of a reorganization plan to Larry. At least this is what I've surmised based on the organization structure presented.

"I want you to keep calm when I go over this and I need your input. Again, this is confidential and I'm only involving you from our group. This isn't final and we were asked to review with any comments.

"I'll only go over the effect on Engineering. Jennifer will have a new position as VP Plant Operations with all six plant managers reporting to her as well as Engineering—and she will keep the materials group."

"Frank, I knew something like this was in the works and have thought about the consequences. First, we have to maintain relations with Tom Watson's R&D group. Second, moving Engineering down in the organization will reduce our influence with product development substantially ... they listen to us, except for their mentally unstable director, because we have the technical knowledge and the authority to change their preliminary designs based on manufacturability constraints. As part of the materials group we'll have no authority. Third, our priorities will be based strictly on product availability and plant equipment status—Engineering will be turned into a high-level maintenance organization. Fourth, Engineering has control of all new facility, equipment, and product launch budgets, which we will certainly lose. Fifth, facility design and coordination of all new division plant construction is one of our main responsibilities—Engineering will lose control. Jennifer doesn't even know what we do in this area. Sixth, our group has managed the molding suppliers since I've been in this job—this is a technical responsibility and Jennifer will stop our involvement altogether. And last of all, I'll have to find a new job."

"Dan, I appreciate your openness and candor. Can you put your concerns in writing, say, by this time tomorrow and elaborate with specific examples you think will get Larry's attention? This is not set in stone and I have the feeling that Larry will not go along with Jennifer having technical responsibility. Larry is smarter than that, at least I hope so, and I'm more than confident that Janet Weatherbe will not go along from a Human Resources perspective. You and Janet are friends . . . talk to her. As a corporate officer she has significant influence. I want you to keep a cool head. Do not, under any circumstance, openly attack Jennifer or any of her reports . . . No more slamming doors . . . and you do not have to find a new job."

"Okay. I can come up with examples . . . just our responsibilities with the molding suppliers and what we have to do from a technical side to keep our products on the market are enough . . . not to even mention the amount of new products and equipment scale-up we do annually, and the worldwide plant expansions. Larry is more than aware of Engineering's involvement. I'll also stress the additional time of getting her to understand technical requirements and geting her approval on anything we do and the impact on launch schedules . . . Is that it? I'm late for the inhaler component design meeting."

"No, one more thing. I was early for Larry's meeting and while waiting outside the door, I overheard the division HR guy, Ralph, telling Larry that their plan to get you on a harassment charge didn't work. What's that about?"

"Frank you don't even want to know."

# 9

# DESIGN FOR MANUFACTURABILITY, DESIGN FOR SIX SIGMA, CONCURRENT DESIGN

## 9.1  PRODUCT DEVELOPMENT MEETING NUMBER 1

**In Attendance**

Keith Carlisle, Program Manager
Susan Jaffey, Quality
Dan Garvey, Engineering
Gordon Taylor, Product Development Design Team Leader
Gary Devoe, Product Development Lead Designer
George Planck, Product Development Designer
Jeffrey Daniels, R&D
Patty Keyser, Marketing

Gordon was the chair for this meeting and did not prepare an agenda. Keith Carlisle noted his disappointment and told Gordon that an agenda was a requirement, with meeting minutes copied to the Device Master Record.

Gordon started, "Sorry, I'll do an agenda in the future. I want to introduce both Gary Devoe and George Planck from Product Development; Gary and George will be the team designers for this program.

*A History of a cGMP Medical Event Investigation*, First Edition. Michael A. Brown.
© 2013 John Wiley & Sons, Inc. Published 2013 by John Wiley & Sons, Inc.

Dan and I went over a few things important for the team to understand in the initial phase of the design process. Dan will take it from here."

"I want to discuss what we need from a manufacturing perspective to assure the components can be produced with the least amount of redesign.

"As you all know the principal goal of our groups is to provide a product that achieves total customer satisfaction and realizes financial success for our company.

"To do this the team must consider the effect of the final component designs on manufacturing—for both our in-house production and that of our molding supplier. For in-house assembly the components must have sufficient tolerance so they fit easily together in the assembly operation; and for the molded components they have to be designed so the molds are simple to operate and maintain and are capable of meeting the specified component design tolerances. In simpler terms, the components must be manufacturable. Manufacturability is defined as the ability to reproduce, identically and with minimum waste, units of product so that the customer's needs are achieved—and so the product meets the program's financial goals."

Dan paused to gauge how the design guys were taking this in, and continued, "Our design team must take the physical and functional requirements into account first; we must examine the quality attributes of each component and set acceptance requirements that are capable of being achieved with little to no rework or waste. If the inhaler has issues in manufacturing we'll not be able to maintain market availability; and last, but most important to our stockholders, we must keep the total manufacturing cost of the inhaler at a minimum.

"Manufacturability begins with concurrent design of the product and the process, not only to meet the functional requirements of the product but also to provide a robust manufacturing process.

"The methodology of concurrent design has been universally adopted as it has proven that the cycle time to gain market entry can be reduced substantially by aligning the product design with the manufacturing process. This is mandatory if our company is to compete and survive in a global market.

"My talk examines the measures used by industry to determine the capability of the manufacturing process.

"The key activities that impact manufacturability can be summarized as follows. Integrated product and process designs are developed, as mentioned, so that the specified design tolerances and process variation are characterized with objective evidence . . . establishing tight component tolerances only increases rejects in production. Final product acceptance requirements must be wider than the sum of the individual component tolerances . . . supplier specifications and designs must be achievable and the raw material specifications open as far as possible to allow for finished goods that permit waste–free manufacturing to occur. I want to point out that waste-free manufacturing is a major goal in our lean manufacturing program. The inhaler component designs must be verifiable and validateable with realistic protocols, and the designs must support process and secondary assembly operations that are qualifiable and account for equipment maintenance and postmarket service.

"To achieve this, the product development group and Engineering must work closely in each phase of the product to include the R&D function, design feasibility, and raw material availability from multiple suppliers. Design control during development, production builds, and postmarket support, and with support of functional areas such as customer service and the marketing and sales organizations, must be included as well.

"There're a number of statistical characteristics that determine a manufacturable product that I'll only reference. The details of the mathematical determination can be spoken to at a later meeting. The basic statistics are known as $C_p$, $P_{pk}$, $C_{pk}$, and DPU, which stands for overall defects per unit produced. Each is a measure of the capability of the process to produce the final product.

"The statistic $C_p$ is a measure of the total specification width divided by the process variation width and determines the capability studies of equipment performance as an evaluation of the repeatability and reliability of the process to produce the product. $C_p$ is used as an acceptance criterion for raw material lots such as the inhaler molded components, to provide information for routine process performance audits, and to establish the effects of equipment adjustments during processing.

"As a rule of thumb, for a $3\sigma$ process a $C_p = 1.00$ indicates the process is capable; a $6\sigma$ process requires a $C_p = 2$, a $C_p$ from 1.00 to 1.33 indicates the process is capable but requires tight control, and a $C_p < 1.00$ indicates the process is not capable.

"The two measurements $P_{pk}$ and $C_{pk}$ are modifications of the capability index to account for specific shifts in production means and are used as acceptance criteria as well. The term 'defects per unit' does not require any explanation.

"Our organizations are based on Six Sigma fundamentals and we seek to produce products that meet Six Sigma production goals. This is not the first time we've designed a new product requiring a plastic housing or interface. But it is the first product with a specific design requiring delivery of a predetermined metered amount of product. So it is even more important we follow the manufacturability guidelines.

"To bring you all up-to-date, Engineering will assist the molder to do complete qualifications on all preproduction molds to assure that the inhalers used in clinical studies will meet our final product design."

Keith interrupted, "I want you to follow a parallel path so the design of the final process equipment, including the molds, is done while the clinical investigations are in progress. This is a risk that we're willing to take to launch this product as soon as possible. And by *we*, I mean Larry."

"As long as the qualifications on the prototype equipment are complete and signed-off by management, and we have a final preliminary design for the final production lines, that will not be an issue. The only risk is if there is an issue in the clinical trials and the prototype has to be redesigned."

Gary asked, "Dan, how does Six Sigma process differ from Six Sigma production?"

"I plan on a two-hour review of Six Sigma fundamentals later in the program. To put it simply, the Six Sigma process focuses on developing and delivering near-perfect products consistently and is a business improvement process that identifies and eliminates mistakes that contribute to defects that ultimately increase our product cost. If an organization follows Six Sigma philosophy, the product will be produced at the highest quality for the lowest possible cost.

"We initially interpreted Six Sigma too literally and found this process to interfere with creativity. We've redefined our thinking and use it only as an aid in our design procedure and as a quality tool, not so much as a method to define our business.

"As an organization we've adopted the philosophy called Design for Six Sigma. We've found that up to 80 percent of all quality problems are design related. Emphasis on the manufacturing side alone focuses on the end of the problem-solving process. Emphasis at the front end— the product design and how the product is manufactured—can eliminate most quality issues. The inhaler design must take into account an understanding of the quality characteristics required for patient use, an FMEA must be conducted, and key variables in the inhaler identified by a procedure called Design of Experiments. Design of Experiments is more technical than I want to cover in this meeting.

"The fundamental premise of the Design for Six Sigma approach is a modification of the DMAIC problem-solving process called DMADV.

"Before I go over DMADV you need to understand the concept of DMAIC.

"In the DMAIC process, the 'D' is the first step in recognizing that there is a problem that must be resolved and a *definition* of the problem scope such that the questions of what, when, where, and the extent of the problem are identified. The 'M' represents the data that must be obtained to *measure* the impact of the problem with a determination of when the problem started and a measure of the extent of the problem. In some books on this topic, the 'D' and 'M' phases are combined . . . but the result is the same. The 'A' is the *analysis* required to determine the root cause of the problem. The analysis phase identifies the potential causes, analyzes existing data, requires a list of verified facts, and compares causes with these facts through a table referred to as a *contradiction matrix*. The contradiction matrix compares the potential causes with the known facts and either eliminates the cause or supports the cause based on the facts. The 'I' is the *improvement* phase, identifying the corrective action to eliminate the root cause, and it includes a determination of the best solution, a prototype of the solution, and testing with objective evidence to verify that the solution is effective. The 'C' is the *control* phase, monitoring the implementation of the corrective action identified in the improvement phase and monitoring the effectiveness of the solution.

"The DMAIC process can be applied to quality improvements, to investigating problems that may surface during the inhaler design and testing, or to determining the source of defects in nonconforming product discovered in the field. For our use as a design tool, replace the word 'problem' with 'product.' Though DMAIC applies, I personally recommend the team follow the DMADV process.

"The DMADV process is a method that can be applied to the creation of a new product. The 'D' represents a definition of the projected product goals and customer requirements. The 'M' stands for measurement and determination of actual customer needs and specifications. This is also referred to as the Voice of the Customer, known as VOC. As part of this step the VOC is translated into design requirements called CTQs, which stands for Critical to Quality. This sounds like a contradiction with the definition stage, but the customer may not really understand the requirements. It may be necessary to change the customer requirements based on available technology, materials, safety, etcetera. The 'A' is the analysis phase to identify the design options that best meet the CTQs. Note that the design options will be compared and rated based on the program goals and is supplemented by application of a fishbone diagram. A major goal is that, whatever option is selected, it must be within the program budget and launch timing constraints. I plan on reviewing the fishbone process in another meeting. The second 'D' represents the development of the design details to meet the customer requirements—or, to be specific with Six Sigma terminology, the Voice of the Customer. And the 'V' is our old friend: verification and validation—verify that the design meets specifications and validate the design meets customer requirements. I want you all to note that this method follows the design control procedures in the Agency's requirements. And after each phase is complete in DMAIV there is a team meeting, referred to as a Tollgate, requiring review and sign-off by the team members before the next phase of the project can begin. In addition, to meet Agency requirements management will be required to review and sign the documents.

"The $6\sigma$ referred to in production is a manufacturing target to have a process capable of meeting a $\pm6\sigma$ acceptance standard based upon the production specification relative to the product specification. In this application $\sigma$ represents the standard deviation of the production lot. The two are universally different in scope but their results are somewhat equivalent and the ultimate goal of the Six Sigma process. A $6\sigma$ production does not happen by chance—it has to be designed into the product and the manufacturing process. To be successful, the acceptance requirements must be fully validated so the final finished-good acceptance specification range is larger than the sum of all the tolerances of the manufactured product—including the incoming raw material and assembly operations."

Dan stopped for a moment to catch his breath and further gauge the audience, "Many of our component designs have specified tolerances as tight as possible without any regard to what the actual tolerance should be. Specifying a tolerance tighter than needed is detrimental to our entire manufacturing process and makes it impossible to achieve a $6\sigma$ goal. As part of our lean manufacturing goals, our organizations need to reduce the rejects in manufacturing. This can only be accomplished by reviewing our product design specifications and acceptance criteria. This will require a requalification of most of our manufacturing processes: both verifications and validations.

"A $6\sigma$ quality level is measured in defects per million opportunities and represents 3.4 defects per million: a yield of 99.999997 percent. Even with this level of quality there is still a chance of nonconforming product. None of our current production lines come close to this measure as a result of incorrectly specified tolerances on subcomponents ... if we do the design of the inhaler and the manufacturing process correctly, we should be able to meet a $6\sigma$ goal. Does this answer your question?"

Gary responded, "I get the idea and will do some reading. Thanks."

"Any more questions? ... If not, I'll turn the meeting back to Gordon."

Gary asked one more question. "What is the House of Quality all about?"

"Gary, the Six Sigma name is Quality Functional Deployment, or QFD. When put in a diagrammatic form it resembles a house. The basis of this procedure is to compare the customer requirements on the left side of the house, or what the customer wants and the customer's priorities, against the design features, how to measure the product's ability to meet the customer requirements, and technical requirements, which are listed in the ceiling of the house. The right side of the house is a comparison of how the requirements are met by competitive products to determine if there are opportunities for an improved product. The foundation of the house contains the specification target values. The roof of the house contains a matrix describing the interactions between the design features. The body of the house contains a ranking, or a correlation of the design specifications to the customer requirements. The entire process is a rigorous approach requiring considerable effort and time from the program team to complete and maintain over the course of the program. In my opinion it is useful as an initial planning tool to get the design team thinking in the direction of customer satisfaction. Since the Oxy-Fox Inhaler is a transfer product we have the customer requirements and technical assessment. I do not recommend you apply the QFD. The design fishbone diagram will be more than adequate and will provide the design alternatives your team needs to investigate. As an example of QFD, let's see what I can do with the coffee cup." Dan paused to draw on the whiteboard (see Fig. 9.1).

Dan continued, "The house is built as I just described. I've taken only a few of the customer requirements and design features to illustrate the procedure. The ratings are values assigned based on the customer priorities. Note that temperature and spillage are most important and both have an overall ranking of 5. The rating is the ranking times the design feature. Thermal conductivity is rated as 5 times 5 and spillage is 5 times 5 as well; cost is not a priority in this example, with a rank of 2. The overall rating for each is the sum of the individual ratings, a value of 49 taking cost into account. We know that the design must meet these two customer requirements. Also the cost is of secondary concern. The example neglects the interactions in the roof of the house

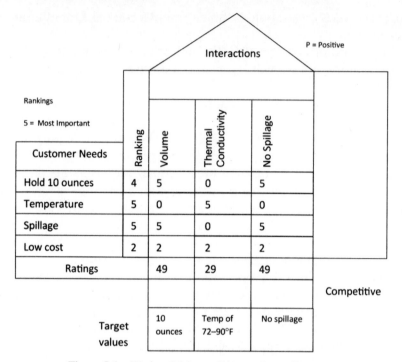

**Figure 9.1**   Design fishbone diagram for coffee cup.

but you can see that both volume and spillage, and volume and temperature, would have negative interactions. The comparative analysis is also not shown since the example does not examine what is on-market and if the new coffee cup offers any improvement. Okay, I can see from Gordon's body language that I am out of time. Need to go to another meeting anyway."

\*\*\*

As this was the first meeting in the program phase, Dan left the design group to attend his staff meeting. He did not like to keep the 30-plus engineers in his group waiting. An important agenda item for each staff meeting is rumors, and Dan wanted to hear everything.

Going back to his office Dan made a mental note to call Janet Weatherbe in Human Resources (HR).

Later in the day, Dan got in touch with Janet and set up a meeting in her office. Janet's office was in the corporate building not far from his building and he walked the short distance organizing his thoughts on how to approach the reorganization question. He would play it by ear.

Janet was in her office waiting for him and invited Dan in to sit down.

"And what do I owe the pleasure of your company? Haven't even seen you for over a month."

"Janet I'll get right to the point."

Janet responded, "You usually do. What's up?"

"Frank shared the reorganization plan with me this afternoon and I just wanted your take on the situation."

"Dan, I have very specific thoughts on this and what we say in this office will stay in this office. Do you understand?"

Dan answered, "Of course."

"Jennifer has been politicking with Larry for some time to get control of all production and, as you obviously know, engineering. Again what I'm going to share is to be kept between us . . . don't even tell Frank."

Janet continued, "We've been friends for as long as I've been with Kinnen and I like you personally. You've qualities that our company needs . . . but you have to learn to keep your mouth shut. This is tough for you . . . so keep your mouth shut on this one!

"Jennifer is out for Jennifer. She's competent in what she does but her methods are unacceptable to me. And personally, I find her attitude rather annoying.

"Larry is no fool. He knows that Jennifer can't handle Engineering's technical responsibility. We base management assignments on acquired skill sets and HR has done a good job of incorporating these into both developmental plans and annual pay increases. This is important in maintaining Kinnen as a market leader in our small but expanding product line.

"To get right to the point, as you do so often, there is *no* possibility that Engineering will report to Jennifer. In fact, another scenario of the reorganization is to have Product Development report to Engineering. I'm sure that Frank didn't even mention it because Watson wants the development group to report to you. Larry is totally against it, but we will see. Tom Watson is a powerful player but sometimes he has the same 'open mouth before thinking' issue that you do. This may hurt Tom in the long run, but for now he's in complete control of the scientific direction of Pharma and carries quite a lot of influence. I've voiced my opinion that this scenario would be best for the organization. Product Development needs a more technical approach to product design. This is what you do and, speaking frankly, they haven't done a good job in the manufacturability area and we've had too many field issues. Ed Chase is a chemist, with no experience in product development, and as you know somewhat mentally disturbed. Their work is in the design of the plastic components of our products, not the chemistries. This is not one of his skill sets or even interests. You're a design engineer with considerable success in redesigning the plastics when issues have happened. You would do a better job.

"Have I answered your concerns?"

"More than I could imagine."

Janet continues, "Enough of business. Any plans for dinner tonight? I'm available if you are. We can reminisce about old times in Pharma. You haven't even *noticed* that I'm still loosing weight. I work out every day . . . how do I look?" As Janet began to laugh, Dan answered. "You look great."

"Let me call Becky. Don't think the kids have anything going on and I can use the break. Where do you want to go and who picks up the check . . . Ms. Corporate VP?"

## 9.2 UPDATE MEETING WITH ED CHASE AND GORDON TAYLOR

Gordon had a scheduled meeting with Ed and arrived on time. The door to Ed's office was open and Gordon knocked. Ed didn't respond and Gordon knocked again. Ed's desk faces the wall with his back to the door. Ed turned and seeing Gordon yelled at him, "Can't you see I'm busy. You should know never to interrupt."

Gordon answered, "Ed, we have a one-on-one. I'm sure it's on your schedule."

Chase looked at his day planner to confirm, telling Gordon to make it fast.

Chase, staring at Gordon said, "Okay. What do you want?"

"This meeting is to update you on the design progress for the inhaler."

Gordon went on to review the results of the first design meeting. Ed did not interrupt but was starting to get visibly flustered.

Chase finally lost control and went after Gordon, "I do not want you to have Garvey involved in our work. You know this. And why do you continually go out of your way to ignore my direction. I'm the Director of this group and you do what I say or get a new job. Do you understand? Are you just stupid?"

Gordon thought for a moment, "*He called me stupid! That's it.*"

Gordon then answered, "Garvey is a key member of the design team as prescribed in our Six Sigma program. He has input we need incorporated in the inhaler design if it is to follow a $6\sigma$ production plan."

Chase answered, "I could give two cents if it is 6σ or not. What you will do is copy one of the devices that your group has reverse-engineered. I've assured Larry that we'll get the design done earlier than plan. Keep Garvey out of it. I don't want to even hear he's at another meeting. Is this clear?"

"Yes it's clear. Garvey won't be invited to our meetings but I'll keep him up-to-date on what we're doing. You can do what you want ... fire me if you want. Our procedures require Tollgate meetings at the end of each design phase and if Garvey is not involved Engineering will not sign off."

At this point Chase's face started to turn a light shade of blue. "Get out of my office and never come back. My doctor told me not to have meetings like this. You never have disagreed with me ... where'd you get the nerve? ... Now get out."

Gordon got up and left. He had stood up to Chase—it was the first step.

# 10

# DESIGN FISHBONE DIAGRAM

## 10.1 LAUNCH TEAM MEETING NUMBER 6

**In Attendance**

Keith Carlisle, Program Manager
Susan Jaffey, Quality
Dan Garvey, Engineering
Gordon Taylor, Product Development
Dave Stall, Materials Management
Jeffrey Daniels, R&D
Lynn Diehl, Plant Operations
Dr. Gonzales, Medical Affairs
Patty Keyser, Marketing

Keith Carlisle opened the meeting, "Short agenda today.

"Both Medical Affairs and Marketing have provided information on finished Inhaler quantities they need to support both the clinical studies and initial market launch. Dan estimated quantities that the molding supplier can provide for each phase. There's a problem with the output

*A History of a cGMP Medical Event Investigation*, First Edition. Michael A. Brown.
© 2013 John Wiley & Sons, Inc. Published 2013 by John Wiley & Sons, Inc.

from the four-cavity preproduction molds but we can work with Materials Management in running the molds earlier than plan to build inventory and bring the components in under experimental-product paperwork. Once the molds are qualified the parts can be released to manufacturing. This won't be a big deal.

"Gordon, Dan, Sue, and Lynn Diehl have met off-line and put together a total market launch plan including timing for each activity. I've reviewed this with Larry and he is okay. Larry was particularly impressed with the idea to place the final assembly equipment off-site at a supplier. This pulls two months from the project timeline and he's pleased. The Project Charter is complete, with all the required information, and the Verification and Validation plans are in draft form. Things look good. The team is ready to begin the design FMEA[*] and Dan wanted to present some preliminaries."

"Keith, before I begin, can we step outside the room for a moment?"

Keith asked Dan what he had in mind. Dan repeated himself and said, "I want to talk privately."

The two stepped into the hallway and Dan tried to keep calm. "Keith, I told you that I needed to work out if we could do final assembly off-site. Quality has said no to this idea. They want the inhaler assembly here—under our control—then we'll move it as a cost reduction once we're confident there aren't any issues. We'll need the two months that you told Larry could be taken out of the timeline put back."

Keith answered, "But I told Larry that we would do this."

Dan just smiled, "And I told you to keep it to yourself until the details were worked out. It won't happen."

After Dan and Keith came back to the meeting, Dan began, "As Keith said, before the FMEA can be started we need to identify all the components of the inhaler and how they interact. To do this a procedure called a fishbone diagram has proven to be most effective.

"A fishbone diagram—also called a cause-and-effect, 4-M, or Ishikawa diagram—is a useful method to identify the design and constraints required to meet the customer requirements.

"The fishbone procedure can be used to identify the design requirements by displaying the many possible components graphically, showing how various components potentially interact, and it follows brainstorming rules when generating ideas.

---

[*]FMEA = failure mode effects analysis (see Chap. 8).

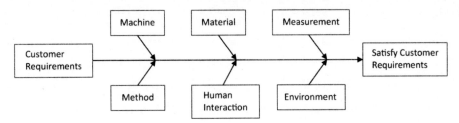

**Figure 10.1**    Fishbone diagram for FMEA.

"A fishbone session is divided into three parts: brainstorming, prioritizing, and development of a design. For our case the 4-M version of the fishbone diagram will suffice. The 'M' in 4-M represents Manpower, Material, Method, and Machine. Additional categories are added as the problem warrants—Measurement, Environment, Human Interaction, Technology, Design, etcetera."

Dan stopped at this point and drew a diagram (Fig. 10.1) on the whiteboard.

Dan continued, "This diagram is the basic approach we should follow in identifying the components of the FMEA and can be used as a basis for the inhaler design. Note the component categories.

"To clarify what the diagram is all about, let's look at the coffee cup example covered in the FMEA presentation" (see Fig. 8.1, Chap. 8). "I've identified elements in each of the 4-M categories that contribute to the design."

Dan put a PowerPoint slide (Fig. 10.2) up on the screen and continued, "Customer requirements are identical in both the FMEA and the fishbone. The difference in the fishbone is the listing of customer requirements and design options that could be included in the cup—ease of use, volume, handle, top, hinged lid, etcetera. The category 'Material' shows a comprehensive list that includes stainless steel, different plastics, elastomers, and a composite. 'Measurement' includes the parameters that will be tested as part of the cup qualification—listed as weight, thermal conductivity, etcetera. 'Method' identifies the types of manufacturing processes, the source of materials, and the safety and design standards. 'Human Interaction' identifies potential failure modes caused through use or through operator error in the cup manufacturing, and 'Environment' identifies potential failure modes such as ambient temperature, cleaning agents that could degrade the selected cup material, road conditions, and vehicle acceleration potentially causing the coffee to spill.

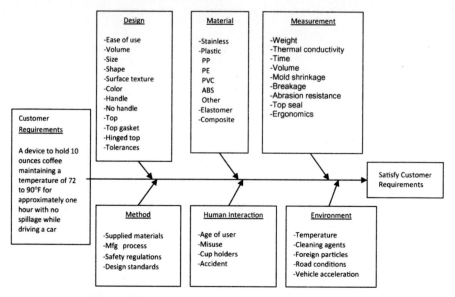

**Figure 10.2**    Fishbone diagram for coffee-cup FMEA.

"From the possible design elements and listed interactions it's possible to identify potential failure modes and modes of control. In the case of the coffee cup, these are not all-inclusive. And, as we'll unfortunately discover, we'll not be able to identify all the design options for the inhaler nor the potential failure modes. It is important to note in the design FMEA example from the last meeting the mode of failure that would cause burns to the user was spillage or leakage and the mode of control was determined to be inadequate to prevent the failure. A new mode of control required a redesign using a hinged lid. Note that, under Design, the hinged lid is listed as an option but was not chosen as part of the initial design, possibly based on cost. Cost does not enter into design when user safety is involved. Any questions?"

Keith Carlisle answered, "Dan good job. Sue, when do you think the FMEAs can begin?"

"Not sure. We have a risk analysis team in Quality that needs to get together. I've enough information to present to them and they'll make the decision."

Keith responded, "Please make this sooner than later."

Gordon asked Sue and Dan if they would give both the FMEA and this presentation at the next design meeting.

Sue answered, "No problem. Include us in the meeting invitation."

Dan nodded his approval.

"Any new business the team needs to address? This could be the beginning of getting these meetings done on time."

No one in the room answered. They all had more than their share of work and needed to leave.

"Okay. That's it."

# 11

# PRODUCT SPECIFICATIONS

## 11.1   PRODUCT DEVELOPMENT MEETING NUMBER 2

**In Attendance**

Keith Carlisle, Program Manager

Susan Jaffey, Quality

Dan Garvey, Engineering

Gordon Taylor, Product Development Design Team Leader

Gary Devoe, Product Development Lead Designer

George Planck, Product Development Designer

Jeffrey Daniels, R&D

Patty Keyser, Marketing

This meeting began on time for the most part with no formal presentation of an agenda or review of minutes from the last design meeting. Not an acceptable way to manage a project. Dan was explaining how the product specifications really follow the Six Sigma methodology.

"Product specifications include the performance parameters which have been defined by R&D and relate to the Voice of the Customer,

*A History of a cGMP Medical Event Investigation*, First Edition. Michael A. Brown.
© 2013 John Wiley & Sons, Inc. Published 2013 by John Wiley & Sons, Inc.

or, in terms of our documentation, the design inputs. From these user requirements this team will design the features of the inhaler: the required features, such as that it must be a metered nasal device; the desired features that marketing may want included, such as the inhaler color and any features that a woman may consider attractive; ease of use for a one-handed operation; the basic inhaler operation and required dosages; storage while in our work-in-process system; and shipping methods. There are additional detailed design drawings for each of the plastic components, and material and assembly specifications that this team will develop over the next few weeks. The most important thing is that we keep in mind the Voice of the Customer, the design inputs, and map of how the design outputs satisfy the design inputs.

"This team will participate in a 'House of Quality' meeting, actually a number of meetings, and prepare a detailed fishbone diagram that will clearly define the design alternatives and their relative weights in meeting the design inputs."

Keith interrupted, "We have not agreed to a detailed House of Quality for this product. You need to review this with me after the meeting."

"Keith, this is a decision that the design group must make. Engineering will participate for manufacturability requirements and that is all. If you and Gordon decide that the 'house' is too much; that is your decision. Remember that we are a Six Sigma organization and our protocol requires us to follow DMADV*."

Gordon interjected, "Before we even get there my group must develop an inhaler design and prepare functional specifications. So far Jeff has provided basic specifications that the drug is an aqueous suspension and must be administered intranasally by an atomizing metering pump. There will be two dosages available: each will have 25 mg oxytocin; one will have 25 mg foxepin; the other will have 50 mg foxepin. Our Patent Department provided two options we could use as a design basis. One device is off patent and we can copy directly, but it does not contain a metering element and is basically an aerosol spray. We could modify the design with a metering component but it would be easier to just start over. The second is the Katlin device, which is still under patent protection. This device might have to be modified to meet the spray volume and accuracy requirements. Anyway, as I said, it is under patent protection. Finance could not negotiate a contract with Katlin to use the device."

---

*DMADV stands for design phases Define, Measure, Analyze, Design, and Verify-and-Validate (see Chap. 9).

Keith asked, "Have you gone over this with Ed? I am sure he will help make this decision."

Gordon thought, "*Yeah, Ed will force us to copy the Katlin device with a slight modification for a different dosage if needed and either get us into a patent infringement suit or make us pay substantial royalties. But in his mind the only goal is that we meet our market launch date.*"

Keith continued, "Okay. Gordon, this is your meeting. As Project Manager I would like to see published minutes and an agenda for the next meeting."

"Sure. Paperwork is my life. I have nothing else to do. Unless there are other issues, the meeting is over."

Dan just listened to the discussion and sensed that the meeting was about over; he stood up, "The meeting is not over. To be consistent with our design philosophy the team must follow the accepted practice in engineering design. First there is a conceptual design phase that follows the Define stage of DMADV, with which we are all familiar. The Concept phase also requires the design team, to gather information on the best design practice by researching technical articles, patents, etcetera; from this research basic design concepts have to be generated and analyzed, with the best concepts taken to the evaluation and final concept with a functional prototype. Once the concept phase is complete, the team moves into specification and fabrication phases. These include descriptions of the product specifications, including materials, detailed engineering drawings to scale with tolerances, and processing methods. The final phase prior to market launch is the product qualifications, which we have covered previously in detail."

"Dan, the meeting is over! I have other things to do and we know how to run a design project. You are going on about the same things that you continuously bring up. The meeting is over."

Keith got up and left, thinking, "*This guy does not have enough to do.*"

# 12

# DESIGN CONTROL

## 12.1   DESIGN TEAM MEETING NUMBER 7

**In Attendance**

Keith Carlisle, Program Manager
Susan Jaffey, Quality
Dan Garvey, Engineering
Gordon Taylor, Product Development
Gary Devoe, Product Development
George Planck, Product Development
Jeffrey Daniels, R&D
Patty Keyser, Marketing

The design group had completed five additional meetings prior to this. At the third meeting Dan Garvey and Susan Jaffey presented a modified failure mode effects analysis (FMEA) presentation with the fishbone analysis. The team appreciated the information and applied the fishbone concept at team meeting number 4 and completed the preliminary component designs during meetings 5 and 6.

*A History of a cGMP Medical Event Investigation*, First Edition. Michael A. Brown.
© 2013 John Wiley & Sons, Inc. Published 2013 by John Wiley & Sons, Inc.

Neither Dan nor Sue had been invited to attend these meetings.
The agenda was short—specific to the final inhaler design.

Gary Devoe began the presentation.

"The design of the inhaler is complete and has been reviewed with management. To save time, we based the design on the Katlin inhaler. Legal has assured us that our design improves the patent and we will not have to pay royalties. The device is simple, with four components not including the drug container. The main body holds the container and is designed to interface with the container cap. The top houses the metering device. Our department director is insistent that only the 25 milligram dosage be provided, with the patient insert telling the user to administer a second spray if the higher dosage is prescribed. As I said, each device is made up of four parts—each specifically designed to control a predetermined quantity of spray."

Gary then connected his laptop to the overhead projector and pulled up the design file detailing each component and the final assembly.

Dan Garvey said, "Neither Sue nor I was included in your last two meetings and this violates our Six Sigma policy. Gordon did send me the team's completed fishbone and FMEA. I've gone over these carefully. In the design portion of the fishbone you've listed the possible alternates for the metering component as 'administered by a pressurized aerosol.' The FMEA addressed potential failures of each design alternate. I noted the container mode of failure was determined to be low pressure resulting in an inadequate delivery of medication or inadequate pump prime. A pressurized device is not even an option: These are for oral inhalation, not nasal. Inadequate pump prime is good, but the mode of control listed 'design.' If the device is not used regularly, the pump will have to be primed. Gordon can you explain this?"

Gordon responded, "Dan we did note in the FMEA meeting that the container would require an additional mode of control. The aerosol version was discarded and the pump prime issue will be noted in the patient insert and drug label. If the device is not used for a period of five days, the pump will have to be primed so the medication delivery would be consistent over the label claim. But we didn't include the change in the FMEA."

Dan continued, "Then you need to update the FMEA with this information. Also, the customer requirement document requires both the 25 and 50 milligram dosages. Has this change been approved by Medical? If so, the document needs to be updated."

Gordon didn't like to be put on the spot and was uncomfortable with Dan's questions but continued. He hadn't told Dan he was ordered by Chase to keep him out of the design and just copy the Katlin inhaler approved by Legal.

"The team decided that the Design mode of control in place was adequate. And the meter component selected is available. It will reduce our design time by four weeks. I'll update the FMEA with the information and I'll get with Medical."

Dan went on to say, "The User Requirements identify a pump system and our prototype equipment is based on this. The modified Katlin inhaler must be tested to guarantee a consistent delivery that will be effectively absorbed by the nasal membrane. I believe the team does not have adequate data to support that each application will contain an equal amount of compound. Anyway, the metering device in the Katlin inhaler does not control the dosage; the orifice in the top does this. You are only modifying the top orifice. Your design does nothing to improve the on the Katlin patent."

Gordon still had the floor and just looked at Dan. Gordon then looked at Keith Carlisle for help.

Keith answered, "Dan the design we have will do the job. It's on the market and is proven. The attorneys have advised that the modifications needed to control the two dosages of medications will allow us to use the design. The design and development of a different delivery system will add three weeks to the program and we'll not have inhalers available for the clinical studies. Changing the design now, as you know, will cause a waterfall effect causing a redesign of the container-filling equipment, which is in-house, and delaying the inhaler assembly equipment we need to support our commitment to Medical Affairs. The design team is satisfied the inhaler will do the job and meet all standards."

Keith responded with obvious annoyance. "Dan it's not your call. Chase has been involved and has approved the design. I've gone over the design with Larry and as long as the Director of Product Development is on board with the design—and by the way this is Ed Chase . . . *not* you . . . Larry is comfortable. It's a done deal. The preliminary design is acceptable. The research group is not concerned. I'll not allow this program to be late. The division is depending on our team to meet the schedule."

Dan then asked Gordon his technical opinion. Gordon answered, "The available system design is good enough. But from a technical concern, I would like to see the device go through complete preclinicals to include toxicity testing prior to the clinical studies. I am concerned that the device will deliver the appropriate dosages. I brought this to the attention of Ed Chase and he threw me out of his office."

Now Dan was getting upset. "Gordon, did you sign off on the design?"

"No, Chase signed for Product Development."

Dan was visibly upset and continued, "As you are aware the final design must be approved by Engineering . . . this is a Tollgate . . . and that signature is mine. I do not care if we miss the goal; it's more important to have a product we can manufacture and meets the safety and purity that the Agency expects. We'll more than likely get through the process qualification and execution of the validation plan. My concern is more in the long run: We're copying a design that requires modification to meet the dosage requirements and we do not have supportive data."

Keith Carlisle just ignored Dan. Dan continued, "You also have to have a lead-in on the meter orifice or the assembly line will never be able to insert the meter into the container cap. And there is no reason to have four parts in the metering device. This can be done with one part and . . ."

Keith stopped Dan in midsentence. "Dan, that's enough. The design is done and approved. Your job in regard to the design is to coordinate with the molders and get the preproduction tooling built. If you have problems, take it up with your boss. The item is closed. The next agenda item is . . ."

Dan didn't stick around long enough to hear the next item on the agenda. He was out of the room and on his way to talk with Frank Hamilton.

As Dan walked to his office, he thought, *"Gordon has kept me out of these meetings . . . don't understand why . . . but is passing the designs to me . . . doesn't make sense. I'm amazed Gordon is standing up for himself . . . this is good. He has to express his expert opinion and not just go along with Keith's push to get this program done on time. I mean, we could be putting patients at risk . . . just to save time in the program . . . But why haven't I been included in the past meetings? And I don't understand why Jeffrey Daniels is going along. Jeff knows that the Oxy-Fox must dispense a consistent dosage with each application for absorption in the nasal passage . . . a reduced or increased delivery pressure was* **identified** *in the product FMEA. Jeff is not standing up to Keith and is letting him make decisions that are Jeff's responsibility. And why is Sue keeping her mouth shut. Something has to be going on with Product Development, Quality, and R&D."*

## 12.2  PRODUCT DEVELOPMENT STAFF MEETING

Ed Chase's direct reports attend this meeting weekly. Also attending for the first time, on Dan Garvey's advice, is Janet Weatherbe, Corporate VP Human Resources.

Chase came in five minutes late.

Without looking around the room, Chase started into a tirade.

"Gordon . . . you have completely ignored my specific direction to keep Garvey out of your meetings. Larry is beside himself with Garvey's insistence to test the inhaler your team redesigned. You know I approved this design. And you are taking sides with Garvey for additional testing. Are you too stupid to understand that testing the design will delay the program? I'm taking you off as lead scientist. Gary Devoe will take over. He at least will follow direction. Who do think you are! How much stupider can you get!"

Gordon answered, "I think I'm the leader of the design team responsible to provide a safe and efficacious product. That's who I think I am. And I've had it with you calling me stupid!"

Chase freaked, turning blue, raising his right arm in a Nazi-type salute and yelled, "Heil Gordon! Heil Gordon!"

Janet was sitting in the back of the room taking it all in.

Chase was so absorbed in himself he hadn't even noticed the Human Resources VP now had first-hand evidence of the man's mental state and the level of abuse subjected on his staff.

Following the meeting, Janet went to Gordon's office. "Gordon, can I have a moment of your time? I want to apologize for the harassment you were subjected to and let me assure you it will never happen again."

Janet went back to her office to talk with Larry and her boss, D.G. Marckam, Executive Vice President of Human Resources.

Gordon was proud of himself. He had stood up to Chase and, for once in his career at Kinnen, he felt good.

## 12.3   ENGINEERING ONE-ON-ONE

### In Attendance

Frank Hamilton, Director, Engineering
Dan Garvey, Group Manager, Engineering

It took over two weeks for Dan to meet with Frank; he explained the situation. Frank had already been told of the incident and had spoken with Larry. Frank just asked, "Does this affect the manufacturability of the inhaler?"

"Frank, that's not the issue. I would like the nasal inhaler completely tested before I approve the design. Using the redesigned Katlin device without extensive testing is asking for trouble."

"Dan, I asked you about the manufacturability. Larry is on my case that you're stepping way out of your area of responsibility. Can Product Development's design be manufactured?"

"It should be optimized to reduce the amount of components."

"Dan, you're avoiding my question."

Dan sat silent for a few moments, finally answering, "Yes it can be manufactured. The metering component will have to be changed to include a lead-in, and I already included the change in the mold design."

"Then you have to approve the design."

"Frank, I can't do that. I have a strong feeling on using products designed for other applications without complete testing, and no testing has been done. I realize that human testing will be part of the clinical studies but, as you know, the Agency has approved abbreviated clincals and there may not be an adequate patient base to see any issues. Anyway, clinicals aren't to be used to test design concepts."

"Larry has directed me to approve for you. Normally I wouldn't do this but in this case I may. If I don't, Larry will sign for Engineering. If I go against him, we'll definitely be reporting to Jennifer."

"Whatever . . ."

The meeting was over. As Dan left his director's office he thought to himself, "*Even Frank is worried about his bonus.*"

## 12.4 PROGRAM UPDATE

### In Attendance

Larry Fletcher, Pharma VP Operations
Keith Carlisle, Program Manager

"Larry we're on schedule. The Oxy-Fox conjugation process, container filling-and-capping equipment, and prototype assembly equipment are installed, and the Installation–Operational Qualification runs are complete. The inhaler design is finalized and the molding supplier has completed the preproduction molds. Garvey is in the middle of mold qualification. Even though Chase approved the single inhaler use, Chase never cleared the change with Medical and we had to return to the two dosage product."

Larry said, "Not important at this stage. One container or two. I couldn't care less. When will inhalers be available to begin clinicals?"

"We released inhalers to Medical Affairs this morning."

Keith paused for a pat on the back. Larry smiled and Keith continued, "There's one problem that you need to address with Frank Hamilton ... You know Garvey has refused to sign off on the design of the components. He's insisting on complete testing before we go to clinicals. Product Development saved us two weeks and the molds are built. If we do the testing he wants, we'll have to add another three weeks past the scheduled design completion date and may have to modify the container filling equipment and who knows how long to change the assembly line before clinical studies can begin. Inhalers were released this morning and Medical Affairs set up specific dates with the clinical investigators ... if we make a change now, the OB-Gyno labs will be on to something else. I'm also concerned with Susan Jaffey ... she hasn't said anything but Sue will probably go along with Dan, and Marcia may bring this whole design issue up in the management review"

"What does Ed Chase say?"

"Ed has approved the design."

Larry told Keith, "This is neither Dan's nor Susan's call. Dan is responsible for manufacturability, and if the design is qualified the Quality organization can't have an issue. Quality's role is review and approval of the documents, not design. If Dan has an issue with manufacturability we can address it. Design responsibility is in Ed's camp. I'll ask Frank to sign in Dan's place. If he doesn't, I'll sign for Engineering. And don't worry about Marcia ... she reports to me."

Larry was happy with the progress and said, "Keith you've done an outstanding job in pulling this all together. Were the validations complete and signed-off prior to releasing inhalers to the clinical investigators?"

Keith hesitated long enough for Larry to ask the question again. "Keith, were validations complete and signed-off prior to releasing inhalers to the clinical investigators?"

Keith finally answered, "As I said, the Installation–Operational Qualification runs are complete and the documentation is being reviewed by Quality. The Process Qualification runs are complete and the functional testing is underway. The molding qualification runs are complete and the data should be reviewed and approved by the end of the week. I released the inhalers from the prototype processes and assembly on my authority. If I didn't, we wouldn't have met the schedule."

Larry was somewhat upset but just answered, "No matter. The important point is you made the deadline to begin clinicals. Unless there's a problem with the functional testing no one will ever know."

# 13

# DESIGN OF EXPERIMENTS (DOE)

## 13.1 MOLDING TEAM MEETING

### In Attendance

Dan Garvey, Pharma Engineering Group Manager
Tim Olsen, Pharma Engineering Molding Section Manager
Mike Muller, Pharma Molding Engineer
Barry Stevens, Molding Supplier Process Development
Engineer

Dan opened the meeting. "The team has made considerable progress in the past ten weeks. All the component molds are fabricated and we're in the qualification phase. I know we're trying to meet a 6σ capability index but with the specified tolerances it isn't possible. The design group specified part dimensions tighter than needed without verifying the tighter tolerances were actually required. No surprise . . . but disappointing. Though our internal acceptance criterion is $Cp = 1.33$, the initial data for all but one part meets the target capability index of $Cp\sim1.0$ for a 3σ process. A 3σ process will have to be good enough. We

*A History of a cGMP Medical Event Investigation*, First Edition. Michael A. Brown.
© 2013 John Wiley & Sons, Inc. Published 2013 by John Wiley & Sons, Inc.

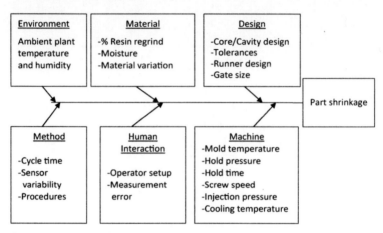

**Figure 13.1**   Fishbone diagram for part shrinkage.

need to find out why this one component is not capable . . . actually, the orifice dimension is measuring only 64 percent at nominal with the remainder on the low side but none below the minimum specified dimension. Tim, I know that you're in the process of experiments. Can you go over what you're doing with the team?"

"The problem is the orifice size on the meter interface. It's consistently on the low side and we think it is due to shrinkage. We've put together a Design of Experiments . . . a DOE . . . to determine the cause."

Tim put a fishbone diagram (Fig. 13.1) overhead on the screen.

"The overhead shows the main factors that can contribute to shrinkage. The environment in the plant—ambient temperature and humidity —could be a cause. The amount of regrind allowed in the plastic resin, moisture, and material variation can contribute as well. The design of the molds themselves—noted as tolerance, runner, and gate size— could be part of the issue. The process variables outlined in the Machine portion of the fishbone affect the shrinkage as well.

"Factors out of our control at this point in the qualifications are ambient plant temperature and humidity . . . the molds are hot-runner tools, so there is no regrind, and lot-to-lot material variation, hopefully, will be kept to a minimum by the resin supplier. Material Supply has negotiated the resin supplier contract with a 95 percent guarantee on resin composition. Since the molds are built, the core and cavity design, the mold tolerances, runner design, gate size, and sensor variability cannot be easily changed. Operator measurement error is something we have to live with. In addition, to keep the cost of the parts down I suggest we maintain the same cycle time.

"The variables that can be controlled—mold temperature, hold pressure, hold time, screw speed, and injection pressure—are our best opportunity in correcting this problem."

Tim looked around the room and continued, "The DOEs we looked into are the full factorial and a modification based on our knowledge of how the molds operate.

"If we did a full factorial, say for 9 variables evaluating 3 parameters, we would have to run $3^9$ experiments, or 19,683 runs, which is unmanageable. If we reduce the runs to evaluate 2 parameters, we still would have $2^9$ experiments, or 512 runs, which is unmanageable as well. The team selected 3 variables with 2 parameters with $2^3$, or 8, experiments. At \$2000 per run the cost to do the 8 experiments is reasonable—a whole lot less than the full factorial at 19,000 runs."

Dan Garvey interrupted, "Tim, we have money identified in the budget to support up to 16 runs so 8 runs is perfect."

Tim continued, "We chose mold temperature, injection pressure, and hold time as our variables, with two parameters." Tim replaced the fishbone file with the experimental plan (Fig. 13.2). "Okay. Now let me explain.

"From our mold process development runs we identified nominal parameter values. The experiment tests the sensitivity to these nominals.

"The experimental plan was run with lows and highs. In the table on the screen, the variation in mold temperature is identified as '-' = 225°F

| Experiment No. | Mold Temperature | Injection Pressure | Hold Time | Percentage Improvement |
|---|---|---|---|---|
| 1 | - | - | - | 55 |
| 2 | + | - | - | 77 |
| 3 | - | + | - | 47 |
| 4 | + | + | - | 73 |
| 5 | - | - | + | 56 |
| 6 | + | - | + | 80 |
| 7 | - | + | + | 51 |
| 8 | + | + | + | 73 |
| | | | Average yield at nominal parameters | 64% |

**Figure 13.2**  Experimental plan.

and '+' = 275°F, with a nominal of 250°F. The variation in injection pressure is identified as '-' = 1500 psi and '+' = 2500 psi, with a nominal of 2000 psi. The variation in hold time is identified as '-' = 10 seconds and '+' = 20 seconds, with a nominal of 15 seconds."

Tim stopped for a moment and pointed to the screen. "When the runs were completed we got percentage improvements as noted. By the look on your faces I need to explain this overhead in some detail. The 64 percent indicates the quantity of parts measured at nominal specification. We need a minimum of 80 percent to meet a 3σ target.

"To find the effect of the change in parameters with regard to each variable, we sum the yields when the parameters are high and subtract them from the sum of the yields when the parameters are low, dividing by number of values."

Tim turned, shut off the overhead projector, raised the screen, and wrote the following on the whiteboard:

Mold Temperature Effect:

$$[(77 + 73 + 80 + 73) - (55 + 47 + 56 + 51)]/4 = 23.5$$

Injection Pressure Effect:

$$[(47 + 73 + 51 + 73) - (55 + 77 + 56 + 80)]/4 = -6$$

Hold Time Effect:

$$[(56 + 80 + 51 + 73)] - (55 + 77 + 47 + 73)]/4 = 2$$

"Now, our largest gain will come from increasing the mold temperature. Note that during the four high-temperature experiments, the injection pressure and hold time were twice high and twice low, which indicates these variables have no effect on mold temperature.

"One would expect to see interactions between the three variables, and this can be checked by an interaction analysis. We collected a lot of data so no additional experiments were needed to determine interactions."

Tim put the interactions file (Fig. 13.3) on the screen.

Tim smiled, "I gather the data needs to be explained as well. In the table, T is the mold temperature, P is the injection pressure, and C is hold time . . . the same variables as in the previous overhead. The interactions are noted as TP for temperature–pressure, PC for pressure–hold time, TC for temperature–hold time, and TPC for the interactions of all three variables. Following the same analysis I wrote on the board to determine the effects of the three variables by themselves, the

|  |  |  |  | Interactions | | | |  |
|---|---|---|---|---|---|---|---|---|
| Experiment | T | P | C | TP | PC | TC | TPC | Percentage Improvement |
| 1 | - | - | - | + | + | + | - | 55 |
| 2 | + | - | - | - | + | - | + | 77 |
| 3 | - | + | - | - | - | + | + | 47 |
| 4 | + | + | - | + | - | - | - | 73 |
| 5 | - | - | + | + | - | - | + | 56 |
| 6 | + | - | + | - | - | + | - | 80 |
| 7 | - | + | + | - | + | - | - | 51 |
| 8 | + | + | + | + | + | + | + | 73 |

**Figure 13.3**    Interactions between variables.

interactions follow." Tim shut off the projector, raised the screen, and wrote the following on the whiteboard:

$$TP = [(55 + 73 + 56 + 73) - (77 + 47 + 80 + 51)]/4 = 0.5$$

"For the interaction of mold temperature and pressure we see the effect has a numerical value of 0.5. This means there is very little interaction, if any. For the sake of time, I'll only write the results of our analysis for the other three potential interactions. The math is identical." Tim wrote the following on the whiteboard:

$$PC = 0 \quad TC = -0.5 \quad TPC = -1.5$$

Tim stopped for a minute to make sure the guys were following. "From the values on the board it's obvious there is minimal interaction effect.

"Note that the process shrink factors are sensitive to high mold temperature, low injection pressure, and high hold time . . . though high hold time is minimal. Since cycle time is affected by hold time, prior to making a change, the impact would have to be investigated further. Lowering the injection pressure could result in 'short shots,' which were seen in the molding process development, so we need to keep injection pressure at nominal. The team selected experiment 6, which yielded an

$80 - 64 = 16$ percent improvement. When we repeated the qualification run at a 15 percent increase in mold temperature, the meter orifice was capable at $Cp = 1.36$. This demonstrates that the process is capable at the $3\sigma$ level. Even with the improvement, I expect the meter parts to come in with a number on the low side of tolerance but within specifications.

"That's about all I have. I expect the qualifications for all the pre-production tools to be complete by the end of the week."

Dan Garvey took over the meeting. "You all have done a good job on this. I can't remember a molding program that has gone as well. Thanks."

# 14

# START-UP ISSUES

## 14.1 OXY-FOX INHALER WRAP-UP AND EQUIPMENT START-UP

The final schedule including all aspects to provide inhalers to the New Drug Application (NDA) clinical studies and the market launch product went fairly well, with only a few issues. The final Installation and Operational Qualification documents including the Process Qualification testing for the finished Oxy-Fox Inhaler passed within the acceptance requirements and were approved by the Quality group. No one was the wiser that the inhalers issued to the clinical investigators were not completely tested and did not have their final documentation approved. Keith Carlisle met his goal in completing the program in less than two years; actually, the program from initiation to final approval for launch was done in twenty months. Keith was happy—not only would he receive a considerable bonus but he had solidified his career at Kinnen. Larry is working with Human Resources to promote Keith to Director Program Management—the second step in his career plan following the MBA.

The determination of the conjugation process parameters and the operation of the Haul-Miller filling-and-capping equipment went

*A History of a cGMP Medical Event Investigation*, First Edition. Michael A. Brown.
© 2013 John Wiley & Sons, Inc. Published 2013 by John Wiley & Sons, Inc.

smoother than expected, with only one failure. The cause was identified as a faulty steam control valve and the corrective action was noted, approved, and implemented in two weeks. The final conjugation system production scale-up did not go as smoothly as the prototype nor did the in-house assembly line—no surprise.

Issues with the final conjugation system were identified early in the start-up and corrective actions were implemented prior to the start of the qualification testing. The corrections were done on a parallel path with the production assembly line and didn't impact the overall program timing. The scientific group determined the changes in the conjugations were insignificant and would not affect product performance: the Quality group accepted and signed off. This was important as the clinical lots were produced on the prototype process with the issues.

Garvey was not included in most of the product development design meetings. In addition to his vocal refusal to accept the Oxy-Fox delivery system without significant testing and the multicomponent meter design that would increase the manufacturing cost, he refused to sign the final documentation. He spoke to Gordon a number of times but Gordon was no longer the lead scientist. He spoke to Gary Devoe, who had taken over the inhaler development, with no success. Dan tried to set up a meeting with Ed Chase but Ed refused to meet. The additional molded components added cost to the product and reduced the overall tolerance budget for final assembly; and the assembly required two additional operations on the assembly equipment, which increased the cost of the equipment and could result in increased production downtime. Most important, the meter orifice didn't have a lead-in to accept the container. The product designer argued it was too late to make the drawing change. Dan worked with the molding supplier to design the final meter with the required lead-in. The component change required the design documents to be modified with objective evidence to support the change. Dan's group provided the evidence. The product design group was very unhappy that they needed to amend their final approved documentation. The Quality organization completed the Verification and Validation plans using the results of the prototype qualifications. The Risk Analysis documents were completed and approved. Validation of the final Oxy-Fox Inhaler manufacturing line was shown to be equivalent to the prototype validations.

Dr. Vickory and the President of the Pharma Division, R.L. Siegal, met with the Food and Drug Administration's D.C. office presenting authoritative literature documenting that tricyclic intravenous injections were well characterized in human toxicity studies and did not represent a patient risk. Dr. Vickory is respected by the Agency and defended that nasal administration of the drug would have similar clinical effect as the well-documented intramuscular injection. From this meeting Kinnen petitioned the Agency to allow abbreviated clinical studies, which were subsequently approved with very little discussion or investigation by the Agency. Clinicals were done concurrently with the design and installation of the final production line and completed in a little over fourteen months.

The final assembly line rate necessary to meet marketing's initial launch estimate was determined to be sixty parts per minute. The equipment was built by a company in Toronto that specialized in this type of assembly—also the only one to bid with a guaranteed delivery per the project schedule. The supplier was not Engineering's first choice. The equipment was a custom design and Engineering had not worked with the Toronto company previously. It would be a risk but Garvey was willing to take the risk to meet the schedule. The engineers in his group would work closely with the Toronto machine builder and assure that the equipment would work properly and be built within budget and time constraints. Initial qualifications—to include three material lots and a minimum of three production runs—would be done in the Toronto shop prior to acceptance at Kinnen.

Marketing insisted on an inhaler body more suitable for a woman to hold than the round configuration designed by Product Development. The change incorporated a recess in the body of the device near the top of the inhaler to allow the user a firmer grip and better control of the container actuation mechanism. It was a last-minute decision and the molding supplier would have had to modify the preproduction molds after they were completely qualified. The design change on the surface didn't appear to interfere with the inhaler function but did take up a portion of the allowed tolerance in the container–meter interface. Dan met with Maria Sanchez, Pharma Marketing Director, and explained the potential risk of the change and why the qualification packages would have to be redone. Maria overruled Patty based on the increase in the program schedule the change would cause, and the change was not made. Maria counseled Patty that in the future she had to bring design concepts to the team early in the program.

When the production assembly line was in place and the Installation Qualification complete, the first Operational Qualification runs began. The line was not capable of running for more than a minute without a jam at the container–meter insertion station. After a number of trials, Lynn Diehl shut the line down and told Engineering that her group was losing money on the operation and wouldn't run the equipment again until all the bugs were resolved.

Garvey spoke to her, "Lynn, we need to run the equipment as much as possible. My group has a high-speed camera and we need to identify exactly why the jams occur."

Lynn replied, "My decision is based on financial performance and I'll not accept any additional department variances due to my people sitting around watching your engineers fix this line in production. I suggest you remove the equipment to an area under your budget and get this thing fixed."

"Lynn, you don't understand. First, there's money identified in the budget for initial manufacturing expense, noted in the budget line item as IME. This money is there to reimburse your department for any variance you encounter. You need to meet with the plant financial analysts and work it out. Second, if you don't run the line, we'll never get the bugs out and we will not meet our market launch date."

"Dan, that's your problem. I have no idea what IME is. Get this equipment out of my area."

Dan was getting nowhere with Lynn and headed for the plant manager's office.

"Sorry to barge in on you but I need you to intervene with Lynn." Dan explained the situation but went further in his description than he should have.

"Bruce, we've known each other for a long time. My group has always supported your plant as a first priority. Lynn has no operating experience at all. You need to get this woman into another position . . . how did she even get the job . . . she's dumb as a rock!"

With these comments Bruce's face turned a bright red and he responded, "I'll look into it."

"That's not good enough. I suggest you move the start-up under Bill Janis. Bill is your most experienced manager and he'll allow us to work this out."

"I said, I'll look into it. That's the best I can do."

*Dan left thinking there was more to this . . . Bruce's reaction was totally out of character. Something is going on. He now reports to Jennifer . . . maybe he has to ask her to make the call . . . but that makes no sense. We've worked together for years. Bruce knows how to run a plant and has considerable experience with new product launches.*

Dan called a meeting with the two start-up engineers in his department. "Okay, what's going on with this equipment?"

Paul Fisher, the lead engineer on the team, answered. "The assembly pick-and-place may not have the tolerance to insert the assembled meter section into the cylinder or there may be a software bug. The unit jams about every twenty operations. We think there may be an issue with the servomotor homing program that is supposed to return the unit back to a zero position. Following each operation the pick-and-place could be returning to a position one degree from center. Our controls engineer, Frank Truk, is looking into the program now. It may take him a few days to figure this out. If not, we'll have to replace it with a different unit with a tighter tolerance."

"Makes sense, get the new unit on-order now just in case. If we don't need it, it can be used as a spare part. Have you looked at the meter orifice dimensions and lead-in on the components you're using? Any other issues with the line?"

Paul answered, "The meter was sampled in incoming quality and the acceptance testing showed the orifice is within specification . . . though on the low side but still within acceptance level. Not sure about any other issues. Haven't run it enough . . . Lynn shut us down earlier today."

"Yeah, I know. I spoke with Bruce and he'll look into replacing her."

Paul kind of smiled, shaking his head. "Boss you do know how to get yourself in trouble. Bruce and Lynn . . . don't quite know how to say this, but . . . are seeing each other after work, so to say. Everyone in operations is talking."

"You've got to be kidding. This woman is not only dumb as a rock but looks like one as well. Ever see a nose that long? She looks like 'Jug Head' from the old 'Archie' comic book. And Bruce is married!"

Paul was still smiling, almost to the point of laughter, "I was kind of fond of Archie's girlfriend Veronica."

"Me too."


\*\*\*

After explaining, with some embarrassment, to Frank Hamilton what happened with Lynn and Bruce, Frank had a talk with Larry. It took two days to get the line moved under Janis's responsibility. Janis cooperated, knowing full well that the manufacturing variances were covered in the project budget. Janis allowed Engineering to completely debug the equipment with the use of his operating staff. Ten runs were made at 1/2 hour durations with complete data on each of the assembly stations. A statistical evaluation pointed to a recurring offset in the pick-in-place device, indicating a software problem. The necessary software modifications to fix the homing device were identified, the corrective action approved and tested, and the addendum included in the Installation Qualification and Validation Master Plan. The line was fully qualified three weeks later than Engineering's schedule as a result of an inordinate amount of time in approving the final documents but met the goals of the overall launch schedule. Dan had allowed for a time slippage of six weeks to correct failures in the prototype equipment and an additional six weeks with the final production line—and used almost every minute. Dan did not advertise the additional six weeks he put in his schedule—after all, this was sandbagging in the eyes of Keith Carlisle, and Dan didn't want to go through another cross-examination.

Dan had two concerns. His first concern was the long-term equipment reliability due to the meter orifice tolerance issue; if there were variations too far on the low side, the assembly equipment could jam. The jam would cause the assembly equipment to stop and would be detected in manufacturing; nonconforming product wouldn't get to the field so there would be no patient impact. Dan's second concern was the delivery system could result in an initial patient overmedication. Dan was convinced that the positive-displacement pump had too many components in the inhaler delivery mechanism and the tolerances on final assembly could result in a larger dosage. He had relied on Gordon and Jeff to intervene in the design but Gordon was taken off the program and Jeff's "stand back and watch" personality allowed Keith Carlisle to make the technical decision for him. Dan had not approved the final design but was overruled by his boss, who signed in his place.

The program goals were met and the first lot to stock was ready for shipment, awaiting Agency NDA approval. After all, the really important result was that upper management would receive their bonus.

Dan had lost Bruce's support and his refusal to sign the inhaler design documents might result in Larry's insistence that he be demoted from Group Manager to an individual contributor. He was confident that his girl buddy, Janet, would intervene and not allow it to happen. Good to have friends in high places.

Jennifer was now Division Plant Operations VP, though Engineering still reported directly to Larry. A surprise in the reorganization, at least to Larry, was that Product Development was moved from Larry's Operations group to report to Pharma R&D under Tom Watson. It was a good decision to provide more technical direction but would move them even further from including manufacturability into the plastic designs.

Ed Chase was removed from the Director position by Tom Watson and given the choice to find another job within Kinnen or retire. He chose to retire and planned a party at his home. Dan was not invited but Gordon attended.

Dan knew that his comments about Lynn would come back to him in time but it was water under the bridge; he had to focus on keeping his mouth shut. The final assembly line was not running at standard and the container filling-and-capping equipment wasn't capable of meeting manufacturing tolerance. Dan needed to put together a continuous improvement team to get these issues resolved. Bill Janis had used all the available IME to cover the line's start-up variances and the plant manager was getting on his case. No surprise. Bruce was no longer a friend.

## 14.2    THE FINAL MANAGEMENT REVIEW

### In Attendance

Dr. Thomas Watson, Pharma VP Research and Development
Dr. Susan Vickory, Corporate VP Medical Affairs
Gail Strom PhD, Pharma VP Quality and Compliance
Larry Fletcher, Pharma VP Operations
Jennifer Feddler, Pharma VP Plant Operations
Marcia Hines, Pharma Director Quality
Frank Hamilton, Pharma Director Engineering
Keith Carlisle, Program Manager

### Not Invited

Representation from Human Resources

Larry Fletcher chaired this meeting. He opened with a brief statement congratulating the group on the successful launch of the Oxy-Fox Inhaler. He then turned the meeting over to his Project Manager, Keith Carlisle.

Keith presented an agenda covering the status of the program, outlining the major timelines with the actual completion dates. The meeting initially went well and some of the attendees were patting themselves on the back for a successful product launch—after all their bonuses depended on the success.

Marcia Hines raised the question of the inhaler design. She heard from Susan Jaffey that the inhaler was modified from the Katlin device and not tested to the satisfaction of the Quality group. Marcia expressed a legitimate concern and wanted to review the decision in detail.

Keith just looked at Larry. Keith didn't want to go up against Marcia. Larry took over.

"Marcia, the Katlin inhaler was chosen to reduce the timing of the overall schedule, and Keith with the assistance of both Product

Development and R&D reviewed the decision in great detail. This is the path the team chose and I support their decision."

Marcia responded, "I would like to see objective evidence in the Device Master Record, and by this I mean data, that the Katlin inhaler is efficacious in the delivery of the metered Oxy-Fox compound to the nasal passage for each dose."

Marcia then asked Gail Strom and Tom Watson if they had reviewed the documentation—neither had.

Larry continued, "I've reviewed the entire situation with Siegal, Pharma President as you know, and he's confident the Agency will accept the NDA with this inhaler. That's all that matters."

Larry then asked Dr. Vickory to comment.

"From a purely medical point of view it doesn't represent an issue. The delivery system is not the primary concern . . . only the quantity of the delivered compound on each successive operation of the device matters. I've reviewed the qualification documents with full functional testing and the validation package, by the way . . . all of which your organizations approved . . . and both show with objective evidence that the final inhaler device meets our design inputs for intended use. The clinical studies are complete, again showing complete efficacy with no aberrant patient reaction, which closes out the data requirements in the Validation."

Keith Carlisle continued the meeting, "Marcia, does this address your concerns?"

Marcia responded, "For the time being. I need to meet with Gail and review the documents from a Quality perspective."

Keith answered, "Marcia this was already done. Quality has fully approved everything."

At this point Larry took over the meeting. "The objective of this meeting is to review the final product launch . . . not to bring up dead issues. The inhaler is fully qualified and the NDA submission is complete . . . that is all you need to know. I'm sure that only minor issues will surface in the field and our group can fix them at that time."

*Both Marcia and Gail held their tongues. This was a no-win situation and only time would tell. They were tired of hearing "fix it in the field."*

Keith continued, "If there are no more discussions, we can circulate the final approval package for management signatures."

Each Vice President signed their respective lines on the final approval document. Gail Strom had reservations but was only one voice and outnumbered. Gail signed the final approval.

Larry took the signed product completion document and just said, "Okay . . . we're done. You should all be proud that the Pharma Division has met a major goal in Kinnen's growth plan."

With the successful product launch an off-site party was scheduled. Most of the main players in the team attended. The party was a gala event—Engineering was not invited.

\*\*\*

Gordon attended Ed Chase's retirement party at his home. Ed had three bars set up so no one would have to wait for a drink and the catered food was excellent. There was even a three-piece band playing near the backyard swimming pool. The party went on late into the night. From the goings-on Gordon understood how Chase's mental state may have developed.

Early in the evening when people were just arriving Chase's wife was cordial and the perfect hostess. After a few cocktails, she began flirting with the young men in attendance, insisting they dance with her . . . this woman is in her early fifties and time had not treated her well . . . probably due to years of alcohol abuse and smoking. Gordon was embarrassed for Chase.

# PART FOUR

## PRESENT DAY: FUNERAL

# 15

## GRIEF

John Tyler's plane was late the day Francesca died; he didn't get home until after nine that night. As his cab entered the driveway he noticed that his mother-in-law's car was parked on the side.

As John entered the house he first saw Sophia in the day room holding little Joseph. He stopped and asked her why she was still there; she normally went home after preparing dinner around six. He didn't notice her tear-stained cheeks. Sophia didn't answer initially and John asked again—he then noticed that she had been crying. He asked, "Is something wrong?" Sophia answered in broken English, much like she did that morning, "Yes, Fran…dead" and started to cry once again. As her body quivered with emotion, she rocked little Joseph.

John, thinking that Sophia had just used the wrong word, asked, "What?" Sophia did not answer—continuing to rock little Joseph, crying more intensely.

At that moment John's mother-in-law, Mary Bucco, came out of the kitchen where she was with Karen and Cathy. Mary came close to John and hugged him hard, burying her head in his coat. "Johnny, I don't know how to say this…Francesca died this morning."

John staggered back and, if not for his mother-in-law's firm hug, would have fallen. He regained his composure, pushing her to arm's length asking, "I don't understand. What're you saying?"

*A History of a cGMP Medical Event Investigation*, First Edition. Michael A. Brown.
© 2013 John Wiley & Sons, Inc. Published 2013 by John Wiley & Sons, Inc.

Hearing their father's voice, the two girls ran from the kitchen, both in tears. They hugged their father, both continuing to cry.

Mary kept calm—she went on to say, "What I've been able to understand from Sophia is that she found Francesca in her bed. She was worried when Francesca came back from her run and went upstairs to see if she was all right. Sophia thought at first that Francesca was asleep but couldn't see her breathing. She tried to wake her but couldn't. Sophia called 911. She was excited and could only speak in Polish. The police later told me that the dispatcher sent police and paramedics to the address displayed through the caller ID system. They came as quickly as they could. Sophia had my number and called me. I couldn't understand what she was trying to tell me but by the tone of her voice I came over immediately. I arrived at the same time the ambulance pulled into the driveway. When we entered the house Sophia was in the foyer holding the baby. She was incoherent . . . she could only point upstairs muttering 'bedroom . . . Fran . . . Fran.' We found Francesca unconscious; there wasn't a heartbeat . . . she wasn't breathing. The paramedics tried to revive her for twenty minutes but she was gone. They even stripped her top and tried to defibrillate. There was nothing they could do. I was in the room with them. I saw how hard they tried. They did all they could. Thank God the girls where in school. I asked that she be taken to Baltimore Presbyterian and called Dr. Gander. Sophia stayed with the baby and I went with the ambulance. He was waiting at the emergency entrance and whatever he could do he did . . . nothing could be done, she had passed."

"But she was fine when I left. I still can't believe this could happen. I don't understand."

Mary Bucco is a strong woman, a woman who is always in control and rarely shows any outward emotion. The next few days would be tough for them all and she had to keep the family together. It was up to her to do what her husband would have done.

Mary gave John Dr. Gander's card with his home number and told him to call.


\*\*\*


Angelo Walden arranged limousine service for the Tyler family over the three days of Francesca's wake and funeral. Mary Bucco was staying with the three children at the Tyler family home. The two girls were taken out of school to be together as a family, and Sophia was with them to help. The entire family was in shock. How could this happen— there was no warning—Francesca was in perfect health. The wake was extended to three days, waiting for the autopsy report. The autopsy would be done at Baltimore Presbyterian Hospital under the direct supervision of Dr. Gander.

The first night of the wake the limo picked up the Tyler family at John's home. Mary was there with Karen and Cathy; Sophia would remain home with little Joseph—the funeral home was no place for the baby.

On arrival at the funeral home, Angelo was waiting. Angelo introduced John and Mary to the funeral director and then the three went into the parlor to view Francesca's remains. John kept the girls seated in the waiting area; he didn't want them to see their mother at that time. This was the home's largest parlor—actually two rooms opened as one to hold the many friends and associates of the Tyler, Bucco, and Walden families. Francesca grew up with the Walden boys and was considered part of the Walden family.

The parlor was laid out with a prayer bench at the foot of the open casket. There were more flower arrangements than the area around the casket could accommodate—spilling out around the perimeter of the room. John and Mary knelt at the foot of the casket and prayed. Angelo stood behind the two. Angelo was accustomed to death; in his business death was sometimes a consequence. But he loved this girl like a daughter and could barely keep his emotions in check.

John and Mary knelt at the casket for a long time. Mary was now holding John tightly with her head pressed against his shoulder.

When the two stood, Angelo pulled John to the side and told him that Anthony and Charles were on their way. Anthony had a small matter of family business that needed attending but he would be there.

Angelo asked, "Johnny, how are you and the family getting along?"

"As well as can be expected, I guess. The girls are taking this hard . . . real hard. It hasn't sunk into me yet. I spoke with her doctor; he will have the autopsy report tomorrow. He needs some time to review the findings with the pathologist and consult with the two staff doctors, and then we'll meet. I trust this guy . . . he'll tell me how this happened. I'll meet with him the day after the funeral."

Angelo continued, "You know you're family. Francesca was like a daughter to me . . . a sister to my two sons. Anything you want . . . ask. And I want to know how she died. I can't accept this. She was too young . . . too healthy . . . she was like my daughter in all respects . . . I need to know what happened."

The next evening more people filled the parlor than it could hold. John greeted all those who paid their respects. Every associate of the

Walden family business attended. Each of them expressed their deepest sympathy…her death was a family matter.

Dr. Summers stopped by to pay her respects. She had taken care of Francesca for the past year. Dr. Summers sat for more than an hour, waiting for John to be alone. She approached him and, following condolences, said, "I want you to send Sophia to my office at 9 AM tomorrow morning. We have the autopsy results and need to get specific information from her. Sophia will have to miss the funeral."

John answered, "That could be tough. She barely speaks English."

"No problem. I've a Polish nurse who can translate. This is extremely important."

The funeral was held at the Catholic church in the old neighborhood. John and Mary sat in the front pew with the children; Mary held little Joseph tightly to her breast. The service went on seemingly forever. Little Joseph didn't cry the entire time, enjoying the comfort and warmth of his grandmother. The mass was followed by interment at the Bucco family plot. The pallbearers consisted of Anthony, Charles, and four Walden associates. Francesca was buried next to her father. Those who attended the interment were silent; only the final prayers of the priest were heard. John, Mary, Karen, Cathy, Angelo, Anthony, and Charles were the last to leave, staying to spread soil on the casket as it was lowered into the ground. John Tyler walked to the limo with the two girls. Angelo, Anthony, and Charles went to their limo as well. Mary stayed a little longer, kneeling by her husband's grave praying. Mary said to Joseph, "take care of our daughter as you did so well when you two were on this earth."

As customary, mourners were invited to the family home for refreshments—only relatives and associates of the three families came. Even though the Tyler home was large there wasn't enough room and John took the overflow to the back garden that Francesca so dearly loved. John noticed the many flowers Francesca helped plant were fading from lack of attention—maybe, he thought, they would die too in sympathy for Francesca. He would have to get on the landscape company.

Anthony pulled John to the side, "Dad says you're meeting with the doctor tomorrow for the autopsy results. When you and Dad meet, I want to be there."

# 16

## THE AUTOPSY RESULTS

John Tyler arrived at Dr. Gander's office early. He sat in the waiting room for almost an hour. Slightly after 10 AM Dr. Gander invited John into his office. The pathologist, Dr. Summers, and Dr. Goodman were also present.

Dr. Goodman, Staff Physician, started the meeting, "John, you have my condolences on the death of your wife."

"Thank you."

Dr. Goodman continued, "We've spent considerable time trying to understand what led to the cause of death ... there appears to be a number of complications that contributed. Officially, Francesca died from cardiac arrest.

"The interview with Sophia yesterday answered most of the questions, and Dr. Gander spent some time going over the potential interactions and causes. As I said, the cause of death is determined but there're complicating factors that led to her heart failure. Dr. Gander will review the autopsy with explanations of our opinion how this happened. We'll try to keep this simple. At any point interrupt and we'll explain. Dr. Gander will have to use some medical terms but he'll explain each detail."

*A History of a cGMP Medical Event Investigation*, First Edition. Michael A. Brown.
© 2013 John Wiley & Sons, Inc. Published 2013 by John Wiley & Sons, Inc.

John said, "Dr. Goodman, I appreciate the time you've all put into this. I'll try to understand but please be patient. This is definitely not my area. Do you mind if I take notes?"

Dr. Gander started the explanation. "John, take as many notes as you like and after the meeting we can do a review and I'll fill in any items you may need clarified.

"As Dr. Goodman said, Francesca passed on from heart failure. We feel this may have been a result of extreme circulatory shock.

"Before I go further, this is Dr. Givens, Staff Pathologist. Dr. Givens did the autopsy and supervised testing of her blood, tissue, stomach contents, and urine. There wasn't much urine in her bladder as it expelled following her death . . . but there was enough for testing. The paramedic report indicated that the bed was actually soaked with both urine and perspiration. This report is key in our analysis.

"The blood analysis contained a very high concentration of an enzyme inhibitor found in licorice . . . this enzyme blocks the naturally occurring conversion of the endocrine hormone cortisol present in healthy individuals to the inactive hormone cortisone. Cortisol is also released in response to a stressful event . . . strenuous exercise is a stressful event, called a stressor. Francesca had been running . . . Sophia told us that Francesca planned to do a five-mile run that morning and she was gone less than 45 minutes . . . and by Francesca's composure when she returned, our assumption is she completed the run. That's an extremely faster pace than she should have done just a month or so after delivery and would account for the high level of cortisol. The high cortisol level by itself is of no concern. What is of concern is the low level of aldosterone, another endocrine hormone, detected in the blood sample. Aldosterone is an important regulator of kidney function.

"Let's review the enzyme inhibitor. In healthy individuals normal cortisol circulation is 1000 times more concentrated than aldosterone. A hormone called 11β-hydroxysteroid dehydrogenase, abbreviated 11β-HSD, converts the active cortisol to inactive cortisone. The inhibitor found in licorice blocks this conversion. This is important because there's a receptor system that controls the amount of circulating aldosterone. This receptor system has a high affinity for cortisol but none for cortisone. If the cortisol is not converted due to the 11β-HSD enzyme, the receptor will be overwhelmed by the cortisol and will not function. The body will not be able to control the amount of aldosterone released by the endocrine system. Without aldosterone, a person will die from circulatory shock as a result of low circulating blood volume. With an excess of aldosterone other symptoms would be apparent that are contradicted in the autopsy report.

"Sophia told us that Francesca had eaten a large quantity of licorice the night before … again, licorice contains the 11β-HSD inhibitor I referred to. This was confirmed from analysis of Francesca's stomach contents, which contained trace amounts of licorice. To repeat, the 11β-HSD inhibitor blocks the cortisol conversion so that the receptors can't recognize the aldosterone level and get very confused, resulting in potential kidney problems.

"John, do you follow me so far?"

John answered, "Somewhat, the kidneys are controlled by aldosterone, and the high level of cortisol interferes with this control."

Dr. Gander continued, "That's correct. But this by itself would not have been fatal. In confusion, the endocrine system normally continues to secrete aldosterone, needed or not, leading to an excess of aldosterone. This isn't fatal but could cause hypertension, edema, hypokalemia … and could eventually lead to heart failure if not corrected … not immediate death.

"But as I mentioned, the blood analysis actually showed a very low amount of aldosterone, which also contradicts the expected response … leading us to suspect that the endocrine system actually stopped producing aldosterone, which was confirmed by a very high level of potassium ion concentration found in the blood sample. The potassium ion retention and sodium ion depletion, caused by the extreme perspiration, could result in a cardiac rhythm leading to arrhythmia and reduced circulating blood volume. These two incidents cause low blood pressure."

Dr. Gander stopped at this point and asked, "John, do you follow so far?"

"Not exactly, but I get the general idea."

Dr. Gander continued, "It's our opinion the arrhythmia was aggravated as a result of excessive loss of plasma volume resulting from the amount of urine and perspiration estimated in the paramedic report."

John interrupted at this point. "I understand so far but what actually caused the heart to fail?"

"I'm getting to that. Sophia also told us that she saw Francesca use her Oxy-Fox Inhaler just after Sophia came to your home. The inhaler contains the tricyclic foxepin. Dr. Gander assures us that Francesca would have taken this medication as soon as she got up. An unusually higher concentration of foxepin than the prescribed dosage was found in the blood sample, so we are assuming that she took a double dose that morning. In fact, Dr. Gander prescribed a 25-milligram dosage and the blood analysis showed a level almost three times that amount. Tricyclics are normally administered orally and must pass through the liver; it could take considerable time before any effect is seen . . . as this is inhaled through the nasal passage the drug enters the blood-stream much quicker and the effect is almost immediate. This is the only explanation the four of us can come up with that could contribute to her death. The extremely high level of foxepin cannot be accounted for.

"From Sophia's interview, Francesca had all the symptoms of an arrhythmia: dyspnea, or shortness of breath; presyncope, or dizziness; and disorientation. With the profuse perspiration she had to be dehydrated. These conditions could lead to a dangerously high heart rate known as tachycardia.

"As mentioned previously the heavy volume of urine expelled and the profuse perspiration caused a dangerously low circulatory blood volume. This low blood volume aggravated the situation, causing the baroreceptor that responds to low arterial blood pressure to excite the sympathetic nervous system, resulting in an unsafe increase in heart-beat. Francesca's heart actually went into tachycardia. To repeat, tachy-cardia is an extremely fast sustained heartbeat. The low blood volume caused a reduction of oxygen to the brain, which sent the lungs into hyperdrive . . . the brain will take control of all organs when it is threatened, and it did . . . The brain must have oxygen to survive. Francesca could not get enough oxygen to the brain, resulting in failure to catch her breath and dizziness. As the brain needs more oxygen the heart is excited even more to increase cardiac output. With a low blood volume the heart has to increase its rate accordingly, eventually leading to ventricular fibrillation.

"Ventricular fibrillation produces uncoordinated quivering of the heart ventricle with no useful contractions. It causes immediate faint-ing, called syncope in medical terms, and death within minutes. Treat-ment is with cardiopulmonary resuscitation, including immediate defibrillation. As you know, the paramedics arrived some time after this episode and tried to resuscitate Francesca but with no success.

"These interactions . . . the 11β-HSD inhibitor consumed in the lico-rice, the low blood volume from perfuse perspiration causing dehydra-tion as well as the high urine excretion, and the actions of the suggested overdose of the tricyclic drug coupled with loss of oxygen to the brain, noted as hypoxia in the autopsy report, sent Francesca into ventricular fibrillation. Blood circulation ceased, resulting in sudden cardiac death.

"John, Francesca died of circulatory shock."

John could not speak. Dr. Gander saw the gleam of tears before John turned his head—he was trying as hard as he could to remain strong. After a minute he said, "This is more than I can bear. I'm afraid I have to leave. Can you put it all in writing?"

Dr. Goodman answered. "As a matter of fact, we will not only give you a copy of the explanation but also plan on submitting a Medical Event report to the FDA's Baltimore district office. The report will automatically trigger an investigation.

"I examined the Kinnen product literature contained with the Oxy-Fox Inhaler and there aren't any warnings of potential interactions with aldosteronism or strenuous exercise, though Dr. Gander did advise Francesca to limit her exercise routine until we became comfortable with the dosage. The Medical Event notification has to be approved by the hospital attorneys before we can send it to the Agency, but this will not be an issue."

Dr. Goodman stopped for a moment to make sure John understood, and then continued, "But what really gets me is the Kinnen product insert states that large dosages of this drug will not cause cardiac arrhythmias. There is no reference that says whether the clinical population included any forms of strenuous exercise or any other potential complicating interactions.

"Also, Dr. Gander was able to get the investigational studies using the hospital's medical internet database on the development of the foxepin compound. It appears that animal studies referenced in the Kinnen product insert did a very low-dosage intravenous injection . . . Dr. Gander couldn't determine if toxicity studies were done at various dose levels exceeding those prescribed for the Oxy-Fox Inhaler. The Agency has expert resources at their disposal who can investigate this further."

As John got up to leave, Dr. Goodman stopped him, saying "John, we need you to bring the inhaler to us."

\*\*\*

Dr. Goodman had prepared a paper for John that put the autopsy report in a readily understandable form. John had then spent most of the evening going over the details and he was able to explain the cause of death when he saw the Waldens the next day.

Angelo and Anthony listened. Both had questions John was able to answer.

The meeting was short; John needed to get back to his office. His life had to get back to normal. Mary Bucco, with continued help from Sophia, was now running the Tyler household.

Angelo and Anthony understood the report. There had to be responsibility. A family life was lost.

Angelo was a member of the South Acres Country Club and served on the club's board of directors with Dr. William Slone, Director of the FDA's Baltimore District Office. Here is where he would start.

# 17

# THE AGENCY

The Agency—the Food and Drug Administration—is headed by a presidentially appointed and Senate-confirmed Commissioner and by several civil-service Deputy Commissioners; there are six Bureaus at headquarters that regulate product areas in Biologics, Medical Devices, Food Safety, Veterinary Medicine, Biotechnology, and Pharmaceutical Drug Evaluation.

The Bureau of Pharmaceutical Drug Evaluation (BPE) is responsible for ensuring the safety and effectiveness of pharmaceutical products and manages the Investigational New Drug (IND) and New Drug Application (NDA) programs for all new drugs. A separate Agency center called the Office of Regulatory Affairs works closely with the BPE and is responsible for assuring products are in compliance with the law and Agency regulations; any noncompliance is identified and corrected; any unsafe or unlawful products are removed from the marketplace.

The BPE Deputy Commissioner is Louis Koppel MD, who is in his early fifties and manages all aspects of BPE responsibilities. He's a diligent civil servant and is well respected by those in the pharmaceutical industry. He tries to keep in close touch with the chief executives of the major healthcare companies and has both informal and formal

*A History of a cGMP Medical Event Investigation*, First Edition. Michael A. Brown.
© 2013 John Wiley & Sons, Inc. Published 2013 by John Wiley & Sons, Inc.

meetings to review pipeline products. The BPE is available for consultation on all new product development and is more than willing to provide early scientific and medical advice. Dr. Koppel has been in the office for the past seven years, maintains a high level of visibility with the Agency Commissioner, and has given presentations at a number of senate inquiries. Dr. Koppel is also an old friend of the Chief Executive Officer of Kinnen Laboratories and has a reputation as being politically connected at the United States Senator level.

There are regional and district offices located throughout the United States—with a regional office in Baltimore, headed by William Slone MD, Regional Director of Pharmaceutical Drug Evaluation and Regulatory Affairs. Dr. Slone holds a very influential position and is determined to assure that the drug companies in his region meet the Agency's Current Good Manufacturing Practice (cGMP) regulations. He's the Agency's recognized expert in Agency regulations and cGMP. Dr. Slone has responsibility for both overseeing the healthcare companies in his region and enforcing compliance regulations. Dr. Slone is a person whom the drug companies do not want to deal with in an audit situation and has previously served as the Principal Investigator on special investigational teams outside his region. Dr. Slone's medical training is in cardiology. Prior to joining the Agency, Dr. Slone had a very successful private practice but gave this up because he felt he could make more of a contribution to human welfare by controlling the quality and efficacy of the drugs used in the care of cardiac patients.

\*\*\*

Cara Williams, administrative assistant to Dr. Slone, took a call from Angelo Walden late in the day. Angelo told her that he was a personal friend and needed to talk with Dr. Slone as soon as he was available. She kept him on hold while she paged the doctor.

Dr. Slone picked up the page and told her to transfer the call. "Angelo . . . good to hear from you. This is the first time you've called my office. Something must be up. Is there a problem with getting the funding for the course renovation?"

"No, the funding is going along better than we ever expected. We've got more than enough funds to do the first nine holes and expect to receive additional donations over the next few months. That is not why I'm calling.

"I don't know if you're aware of the death of my niece, Francesca. She died under questionable circumstances from cardiac arrest. She was healthy with no apparent signs of any cardiac disease. I have an abbreviated autopsy report that explains the circumstances in some-what layman's terms. I would like to meet with you to get your take."

"Angelo, I'm so sorry to hear this, but, how can I help?"

"The doctors who took care of Francesca during her pregnancy and postpartum care feel her medication contributed to her death. They're filing a report with your office. I think it's called a Medical Event."

"I understand. I need a few hours to finish up. Can we meet at the club around eight this evening?"

"See you there."

\*\*\*

Angelo arrived at the club early. He needed a drink to calm down before meeting with Dr. Slone. He sat at a table not far from the bar. The area is decorated as an old Irish Pub serving Guinness and Smith-wick's on tap: a great Irish stout and ale, but he preferred Budweiser. Angelo was the only person in the pub other than the bartender—the place was empty. He was head of the fund-raising committee and thought the board must do something to attract members to spend more time in the bar . . . this was lost money . . . his fund-raising would be easier if the pub could show a profit.

As Angelo nursed the Bud, he reminisced about how he had gained membership four years ago. Acceptance was complicated due to a government investigation of the family waste-management business and required a substantial donation prior to consideration. Angelo has been on the board of directors for over two years and heads the fund-raising committee. Though he had hardly ever played golf he was a natural . . . and he loves every minute of it.

Dr. Slone came in at exactly 8 PM. Promptness was one of his virtues. Angelo watched him enter. Slone is a tall, good-looking man in his mid-forties; his hair is cut short, starting to gray on the sides, and he obviously keeps in shape. Angelo thought Slone actually looked like a doctor and wondered why he left a lucrative practice to serve as an administrator in a civil service position. Someday he would ask.

Dr. Slone sat across the table from Angelo—he motioned to the bartender to bring the same as whatever Angelo had. Since they were the only two in the bar the Budweiser came quite quickly. Dr. Slone looked at the bottle saying, "Angelo, you have to raise your drinking standards," and asked the bartender to bring him a "Smitty's."

"Bill, let met get right to the point. As I said on the phone, my niece, Francesca, passed on last week and I need you to look into it." Angelo gave Slone a copy of the hospital report.

The report was not long and outlined the principal cause of death with the interactions—specifically the potential contribution of the Kinnen Oxy-Fox Inhaler, noting the high concentration of the tricyclic found in the blood analysis and the lack of medical contraindications in the insert. Dr. Slone spent less than a minute reviewing the conclusions. He looked at Angelo and said, "When the Medical Event reaches my office I'll meet with the hospital and get the actual autopsy. I'm not sure which region of the Agency the Kinnen investigation would fall under, but since this death happened in my region I can insist that my office provide the Principal Investigator. And I can assign myself in this role."

Angelo just said, "Thanks, I could ask for nothing more."

The two sat for a while talking club business, the upcoming tournament and who they thought would make the best partners, though partners were assigned randomly.

After Angelo left, Dr. Slone sat for a while reading the report thoroughly. He made a mental note to get the IND and NDA sent to his office first thing in the morning. He also remembered seeing in the Agency Audit Report a number of audits for Kinnen manufacturing plants. He would get those as well.

# PART FIVE

## AGENCY MEDICAL EVENT LETTER

# 18

# KINNEN NOTIFICATION

## 18.1  ANOTHER AGENCY LETTER

As Kinnen's Chief Executive Officer, M.V. Brooks began his day earlier than most. He arose around 6 AM to read the Journal online and watch the morning stock report. When his wife got up and while she prepared breakfast for the two of them and their three teenage children he read the local news thoroughly with specific attention concerning the health-care industry. His wife asked, "Anything important?"

Brooks answered that the industry was down in general but noted since the launch of a number of new products Kinnen stock had gained two points and that would be welcomed by the stockholders, and this was what his job was all about.

Brooks arrived at the office at 7:30 AM and spent time reviewing his schedule for the day. His admin, Carole, would be in promptly at 8:00 AM. His schedule is heavy with meetings the entire day. The first meeting is at 8:30 AM with the Pharma President Siegal and the R&D guy, Watson. He needed an update on the new launches and the two products in research stage that would complement the immunology line. At two that afternoon he had a phone conference with a Wall Street investment firm—he wrote a note to Carole that she should make sure Jameson is available to cover any financial question that might arise.

*A History of a cGMP Medical Event Investigation*, First Edition. Michael A. Brown.
© 2013 John Wiley & Sons, Inc. Published 2013 by John Wiley & Sons, Inc.

His morning went well and Brooks had enough information for the investment firm call. Jameson was onboard with the financials.

At 1:00 PM Carole paged Brooks, telling him she needed to talk to him immediately—a registered letter had arrived from the Agency.

The letter was short, announcing that a Medical Event had been received by the Agency involving the death of a woman and implicating Kinnen's Oxy-Fox Inhaler product. The Agency listed the product identification and lot number and expected Kinnen to conduct an internal investigation with a formal response to the Agency within two weeks from receipt of the registered letter.

Brooks told Carole to get Bob Siegal, Tom Watson, Susan Vickory, Gail Strom, and Larry Fletcher in his office at five that afternoon and went on to meet with Jameson for the call to the investment firm. He thought, "*Another Agency letter . . . no big deal.*"

## 18.2    MEDICAL EVENT REVIEW MEETING

### In Attendance

M.V. Brooks, Kinnen Chief Executive Officer

R.L. Siegal, Pharma President

Dr. Tom Watson, Pharma VP R&D

Dr. Susan Vickory, Pharma VP Medical Affairs

Larry Fletcher, Pharma VP Operations

Gail Strom PhD, Pharma VP Quality and Compliance

Brooks began the meeting, "Thanks for meeting on such short notice. I have a letter from the Agency that you need to be aware of."

Brooks read the letter, pausing a few times to register the reactions of those in attendance.

"Any questions, concerns?"

Gail Strom commented, "We keep file samples of every finished-goods lot. I can get the right people together first thing in the morning and put an investigation team in place. The team will be at the Director level and I'll have Marcia Hines, the Pharma Divisional Quality Director, be the lead. Two weeks may not be adequate time to actually test these samples but the team will do their best. If we could get the actual inhaler the Agency is referencing, it would be helpful."

Brooks answered, "The Agency is keeping the inhaler and doing their own testing."

"That's not good. The implicated product would normally be returned to us. Don't understand why the Agency is doing testing at this point. This is not a formal Agency investigation—at least not at this point; this is an inquiry telling us to do the investigation."

Brooks continued, "I don't understand myself. I need to make a call to the Washington office and find out what's going on. Anyway, put your team in place and follow our standard investigation procedure."

Larry Fletcher, looking directly at Gail, chimed in, "I want our Program Manager Keith Carlisle to lead the investigation. Your groups will provide the support."

Gail met Larry's glare with her steel blue eyes and said without any misunderstanding, "My compliance group has responsibility. Keith can be on the team if you like but Marcia Hines is the lead."

Larry is not accustomed to having his decisions questioned and was visibly upset. No one else in the meeting commented; the investigation is Gail's responsibility and she will be the Agency contact.

Brooks continued, "Larry, this isn't a panic situation . . . don't get excited. Gail is responsible for conducting internal investigations . . . It's in our Quality Policy. We've dealt with the Agency before and I'm not concerned."

Gail interrupted, "Mark . . . yes, that's correct. We've dealt with Medical Events and the Agency has relieved us of responsibility . . . but we've never had a product implicated in a fatality. This will be different. I know you have a political connection that intervened in the past. Do you believe that if the investigation finds any cause resulting from our product that you can keep whoever you know on your team?"

"Gail, first and foremost, I never want to hear you say 'your team'; we're a company . . . this is 'our team.' Got it! Any other discussion?"

Dr. Vickory had kept quiet, wondering whether she should bring up the university studies or not; she made her decision. "We may have a problem. I didn't bring this up previously because our NDA clinical investigations were as perfect as you can get. But on review of the university studies with the tricyclic foxepin the toxicity studies were done with oral dosages of foxepin, not by injection. Foxepin was designed to be taken orally. The data presented to the Agency allowing abbreviated clinical studies was based upon intravenous injections of tricyclics. Though foxepin is a class of tricyclics, it is a new drug and was not tested by injection. As you probably know, the human metabolism can substantially reduce the bioavailability of the free drug prior to absorption in the body's circulation. Our inhaler is absorbed through the nasal passage and possibly 100 percent of the drug could circulate in the blood. Tricyclics have an effect on the neurotransmitters causing an increase in cardiac response. Taken orally the university claims that an overdose will not cause tachycardia. But this claim is only for the oral dosages used in their studies. The Medical Event Report lists the cause of death as ventricular fibrillation, which is usually a result of uncoordinated tachycardia. If the patient had overdosed on the inhaler, it may have contributed to the episode described."

Brooks asked, "So what. Our clinical studies demonstrated the drug was safe and effective as a nasal spray. What's your point?"

"My point is . . . toxicity studies are required by Agency regulations and were not done for either injection or nasal administration. In addition, the clinicals were abbreviated. There may not have been adequate study populations demonstrating all possible interactions or side effects."

\*\*\*

Leaving the meeting, Gail stopped Tom Watson. "Tom, I've been holding my breath since we signed off at the Oxy-Fox Final Management Review. The product launch documentation is in order but if our report to the Agency raises any questions, they'll have an investigation team in here before we can blink. Brooks is overconfident with his political tie. It can only protect this company so far. I especially don't like that the Agency is keeping the inhaler . . . the information Vickory went through is sure to come up."

Gail went back to her office and sent a memo to her admin to set up a meeting with Marcia first thing in the morning and to get Dan Garvey to attend.

# 19

## INVESTIGATION TEAM MANAGEMENT

### 19.1. MORNING MEETING WITH GAIL STROM, MARCIA HINES, AND DAN GARVEY

Garvey came in a little after 8:00 AM—he normally was in at seven but this morning was different. Since his promotion to Director Product Development reporting to Tom Watson, the amount of work required coming up to speed on R&D activities had been keeping him in the office fourteen hours a day. Last night he didn't get home until after midnight. He was tired and had slept in for an additional hour.

Dan's admin stopped him in the hallway before he even got to his office, telling him he needed to get to Gail Strom's office right now.

Dan asked, "Do you know what this is about?"

The admin answered, "No idea."

Dan had to go to the corporate building two blocks from his office. He made this in record time, arriving only fifteen minutes late.

When he knocked on Gail's office, the door was closed and the blinds drawn; Marcia opened the door and invited him in.

"Sorry I'm late."

Gail answered, "No problem. Marcia and I needed some time alone anyway. Dan, before we start I want to congratulate you on the

*A History of a cGMP Medical Event Investigation*, First Edition. Michael A. Brown.
© 2013 John Wiley & Sons, Inc. Published 2013 by John Wiley & Sons, Inc.

promotion. In my opinion this is long overdue. You'll do a good job. Tom Watson is a good guy to work for and I'm sure Janet had an influence."

Dan just said "thanks" and sat down.

"Dan . . . Marcia and I have been reviewing a request from the Agency implicating our Oxy-Fox drug in a fatality. We're required to conduct an internal investigation over the next two weeks."

Gail gave Dan a copy of the letter and instructed him that at this point it was confidential.

"I'm putting together an investigation team and Tom agreed that you'll represent Product Development yourself. You were on the launch team and I'm told you had a number of issues with the drug delivery system. Marcia will lead the investigation but you'll have technical responsibility. Larry has insisted that Keith Carlisle be on the team. I don't want Keith taking charge. You and Marcia must be very careful that the investigation procedures are followed to the letter. I'm certain that we'll not get out of this with just our response . . . there will be a formal Agency investigation. Both Brooks and Siegal are confident their Agency connections at the Washington level will do damage control for us. I'm not confident at all . . . there's been a death and as far as I'm concerned the Bureau of Pharmaceutical Evaluation, or BPE, in the case of a death is relentless in determining the responsibility. And Marcia, you report directly to me until this whole thing is over. Under no circumstance do you take direction from Larry."

# 20

# DMAIC INVESTIGATION PROCESS

Gail put together an investigation team consisting of the following:

Marcia Hines, Pharma Director Quality and Compliance—team leader
Dan Garvey, R&D Director Product Development
Keith Carlisle, Director Pharma Program Management
Paul Fisher, Lead Project Engineer Oxy-Fox equipment
Bruce Garlin, Site Plant Manager
Sue Jaffey, Quality
Dr. Jim Gonzales, Medical Affairs

Marcia opened the meeting. "I assume that you've all been briefed by your management that we received an inquiry from the Agency implicating the Oxy-Fox Inhaler in a patient death. The Agency has requested a formal internal investigation to be complete and returned to them in two weeks; this gives us twelve days. Not a lot of time but the product was launched three months ago and the information should be fresh in your minds and all the qualification testing and documentation is available."

*A History of a cGMP Medical Event Investigation*, First Edition. Michael A. Brown.
© 2013 John Wiley & Sons, Inc. Published 2013 by John Wiley & Sons, Inc.

Keith Carlisle asked, "Marcia, why is this being done at the Director level? I could get it done with the initial team members in a matter of a few days."

Marcia answered, "Keith, our product is implicated in a death. Gail chose the team members. I feel you can answer a number of questions that we'll have; this is the only reason you are even on the team. Questions about why the product was able to launch in record time. Questions about why extensive testing of the modified Katlin inhaler was not done. Get the picture?"

"Marcia, I am sure that Larry would want me to handle this, and don't forget you report to Larry as well. I don't like the tone of your voice and I'll need to talk with Larry, as you should."

"Keith, it is unlikely that I can forget I report to Larry, but this is a quality and compliance investigation. The reporting structure is through Dr. Gail Strom, Corporate VP Quality and Compliance. I've asked Dan to present the DMAIC* method which we'll follow. Dan . . ."

"Keith, to answer your question . . . this investigation is being conducted at the Director level because we have the authority to make the data collection a priority in our respective organizations.

"I'm not going to go over the DMAIC basics . . . you all have been trained in these activities. For this investigation the team will follow a modification of the DMAIC format noted as MAIC. since the Define phase has already been determined by the agency. Basically the two are equivalent. Marcia and I spent considerable time over the last two days outlining the data needed to complete the investigation, with assignments and timelines.

"First, the definition of the problem will be determined in the MAIC Measurement phase. From the Agency letter our Oxy-Fox Inhaler is implicated in a patient death. This is not the problem statement. The cause of death may have nothing to do with the inhaler. It's our job to determine if the inhaler could have potentially contributed or not.

"We'll need data from the qualifications and quality acceptance records for both the incoming material inspection and production areas. In addition, we may need to test retained inhalers from the finished-goods lot implicated in the Agency letter. We've released only five lots: the 25-mg complaint lot, two additional 25-mg lots, and one 50-mg lot. Too bad we don't have the actual device noted in the complaint.

"As part of the Measurement phase the team needs to identify and describe the key components and criteria, identify the information

*DMAIC stands for design phases Define, Measure, Analyze, Improve, and Control (see Chapter 9).

| Topic | IS | IS-NOT |
|---|---|---|
| What | • The pump system.<br>• The dosage delivered over the label claimed applications.<br>• Overfilling of the container.<br>• Manufacturing control issue.<br>• The tolerance stack-up parts in the pump system.<br>• The toxicity of the drug if abused.<br>• Complaint specific to 25-mg inhaler lot.<br>• Or a combination of all the above. | • Complaints could have been received for the next two 25-mg inhaler lots but were not.<br>• Complaints could have been received for the first 50-mg inhaler lot but were not. |
| Where | • Market launch was kept to the eastern sales area.<br>• The death was in Baltimore.<br>• The defect could be in the container or pumping system.<br>• The defect could be a result of a manufacturing issue.<br>• The defect could be patient use. | • Complaints were not received from other cities in the eastern sales area.<br>• The defect could have been seen during final release testing of the four market launch lots and the implicated lot but were not. |
| When | • The Medical Event complaint was dated 12 weeks following market launch. | • Complaints could have been received from the additional lots but were not.<br>• There could have been more complaints but were not. |
| Extent | • To date, only one complaint received. | • Complaints could have been received from all the market release lots but were not.<br>• Additional complaints could have been received from the implicated lot but were not |

**Figure 20.1**  The Is/Is-Not chart used in the investigation.

needed in order to write a problem statement, and provide a background description of the problem. The problem statement is actually the definition of the problem. The tool in our investigation procedure is called the Is/Is-Not Chart." (See Fig. 20.1.) "The Is/Is-Not Chart captures the what, where, when, and extent of the problem by answering questions. The Is/Is-Not questions assume there is a defective product so please bear with me—the answers to these questions can be helpful in determining if there is a defect or not.

"The 'What Is' part asks the questions 'what specific object has the defect?' and 'what is the specific defect?' In our case the specific object is the inhaler. The specific defect, if there is a defect, could be a number of issues. For now, we feel that the specific defects could be the delivery system, the dosage delivered over the label claimed applications, the toxicity of the drug if abused, a specific lot of the inhaler, a potential manufacturing issue, or a combination of these.

"The 'What Is-Not' asks the questions 'what similar objects could have the defect, but don't?' and 'what other specific defects could reasonably be observed, but are not?' This doesn't apply . . . since the inhaler is unique, there are no similar objects that we need to investigate. Answering the second 'What Is-Not' question is difficult. The Medical Event is specific to one lot of product. This was the first lot released. We put together a file, which I'll review on the overhead projector in a moment, that goes over the 'What Is-Not' in more detail.

"The 'Where Is' asks the questions 'where is the defective object observed geographically?', 'where is the defect on the object?', and 'where else is the problem?' The only geographical information we have is the market launch was kept to the eastern sales area and the death was in Baltimore. The analysis will also address 'Where else could the problem be?'

"The 'Where Is-Not' asks the same type questions as the 'What Is-Not': 'where else could the defective object be observed, but is not?', 'where else could the defect be on the object, but is not?', and 'where else could the problem be, but is not?' The market launch was contained in one sales region . . . the complaint was specific to Baltimore. Complaints were not received from other cities in the region. The second question specific to where else the defect could be on the object is tough and will require analysis of the inhaler. The inhaler is simple . . . only the container–cap pump or the meter subassemblies could contribute to the problem, if the problem is actually with the inhaler. In our opinion, the question 'where else could the problem be, but is not?' involves an analysis of our product release testing. For now, the defect could have been seen during final release testing of the market launch lots and the implicated lot, but was not.

"The 'When Is' asks the questions 'when was the defective object first observed?', 'when did the defect level first increase?', 'when, since the level increased, have additional defects been observed, and is there a pattern?', 'has the defective level changed?', and 'when during the product life cycle was the defect first observed?' As for the first two questions, the only information we have is the Agency letter and it references only one Medical Event. There is no other information. Also, our Qualification, and Verification and Validation documents can be used to answer the product life-cycle question.

"I was the Engineering rep on the launch team and am quite familiar with the qualification results. There were no issues and the data was included in the New Drug Application ... the NDA ... to satisfy the Verification and Validation submittals. We need only summarize this data in the investigation."

Marcia interrupted, "Dan, I reviewed all the documents ... please explain to the team why you didn't sign off on the design documents."

"Marcia, I had a fundamental disagreement with the design of the inhaler from a manufacturing perspective, and I felt that the device wasn't tested adequately. The team will review this in the Analysis phase as a potential cause."

Keith Carlisle kept quiet for some time but asked, "Dan, you were the Engineering rep on the market launch and your boss at the time signed off on the design documentation for the inhaler from a manufacturability perspective. That is all that is needed for the investigation. Your personal opinion on the design is not pertinent. The qualification testing, verification and validation testing, and human testing in the NDA confirmed that the inhaler was both safe and effective."

"Keith, you have a good point but the analysis must include the manufacturing information as well as all our testing data. The qualification results may eliminate this as a root cause but that comes later. Not now.

"The 'When Is-Not' asks the complementary questions, 'when else could the defective object first have been observed, but was not?', 'when else could the defective level increase, but did not?', 'what other times could the defective object have been observed, but was not?', 'what other times could the defective level have changed, but did not?', and 'when else in the life cycle could the defect have been observed, but was not?'

"These questions are good. Since the inhaler passed all testing in our qualifications we know that the product was safe as used in our studies as well as the NDA clinicals. We need to answer the questions based on patient usage if possible.

"The last set of questions determines the 'extent' of the problem. These questions ask 'how many objects have the defect?', 'what is the size of a single defect?', 'how many defects are on the object?', and 'what has been the trend?' Since we cannot get the actual device back from the Agency, these answers have to come from the analysis of the retained inhalers from the same lot as the device in the Agency's Medical Event Report."

Dan stopped for a moment to gauge any response, and continued, "The complementary 'Extent Is-Not' questions ask 'how many objects could have the defects, but do not?', 'what other size could the defect be, but is not?', 'how many defects could there be on the object, but are not?', and 'what other trends could have been expected, but were not observed?' Again we have insufficient information on the extent of this problem . . . right now there is only one complaint.

"That's all we need to determine the problem statement. In answering the preceding questions for what, when, where, and the extent of the problem we have a problem definition.

"Marcia and I prepared a file summarizing the information I just went over."

Dan brought up the file (Fig. 20.1) on his laptop and distributed copies to the team.

Dan asked if there are any questions. No one responded. Dan turned the meeting back to Marcia. Marcia took the floor and went over responsibilities for each organization. The Quality organization would do most of the data gathering under her direction.

Marcia continued, "Looking at the overhead it appears that the problem, if there is one, may be in the first 25-mg lot released.

"Once we have information from the lot samples, my group will conduct a severity assessment. In our case, the severity assessment addresses patient risk determined from all the gathered information . . . Much like a risk assessment, which we covered in the FMEA* meetings."

Marcia continued with an explanation of the remainder of the investigation procedure. "Following the determination of the problem statement in the Measurement phase, the next step in our MAIC procedure is 'Analysis'. The team needs to identify potential causes that could cause the problem, analyze existing data from our documentation, construct a list of verified facts pertinent to the development and use of the inhaler, compare the potential causes against the verified facts through the construction of a 'contradiction matrix,' eliminating those causes which are not supported by facts, and collect any additional data that is needed to clarify or verify the problem facts.

"The objective of the analysis is to determine the root cause. The potential causes we've identified so far applicable to the inhaler are dimensions of the metering section, fit of the metering section with the container–cap pump, potential problem in manufacturing, and improper

*FMEA is failure mode effects analysis (see Chap. 8).

use of the inhaler. The final bit of information that really concerns me is the complaint is specific to the one lot."

Dr. Gonzales raised his hand and was acknowledged. "You need to add a potential cause. Dr. Vickory and I met last night to review the university studies used to support the toxicity requirements for the NDA. You need to add that the toxicity studies were done with ranges of oral foxepin administration, not injected foxepin."

Marcia stood silent for a moment not believing what was just said, "That is not good."

She continued, "I'm sure if we had more time the team could identify additional causes. Unless something comes up in your analysis we'll stick with the causes noted . . . including the lack of toxicity information for an injected product.

"I'll have Susan Jaffey get the incoming inspection data for the metering components and the cylinders. Bruce, I need you to pull the file samples with the acceptance records for the implicated inhaler lot as well as the other four released lots, and Sue will also pull the qualification reports. This data will be sufficient to complete the facts section of the analysis.

"Comparing these facts with the potential causes in the contradiction matrix should identify the root cause, if any. Just be sure that the facts are verifiable and note the specific documentation used in the verification. If additional testing is required, let me know immediately and I'll make this a priority in our testing lab. Once the contradiction matrix is complete, we need a formal investigation plan. If the existing data meets all acceptance requirements, a plan will be easy . . . if there's additional testing the team recommends or we find any nonconforming data, the team will have to identify exploratory studies to be included in the investigation plan and these studies will be conducted as a priority. Each study will have a preapproved protocol signed by the group responsible and the Quality rep, Sue Jaffey. Following implementation of the protocol, a management review by the directors in this room will be done prior to further action. Any questions?"

Marcia continued, "On completion of the Analysis phase the Improve phase begins. The Improve phase includes the determination of the best solution to eliminate the root cause identified in the Analysis phase, and a pilot of the solution to verify the solution works . . . with complete qualification documentation. For this study, if we determine there is a problem with the inhaler, the Agency requires the Improve phase just to be in plan form. We don't have time to prototype a solution and the Agency does not expect us to do so in the two-week time requirement. But if a problem arises, the report will include both a timeline to test a pilot to verify that the solution eliminates the root cause, and a timeline to implement the correction in production. Gail will have to report the status of this plan to the Agency monthly.

"Any questions?

"To continue, the last phase of the MAIC procedure is the Control phase. If we actually identify a defect in our inhaler system, the solution will be part of the report to the Agency. A Control plan will also be included to assure that the solution is effective. This plan will be our current acceptance requirements if the Analysis phase determines our testing to be acceptable. If not, new acceptance and testing protocols will be included with an implementation schedule. This status will also be reported to the Agency monthly. That's about it. We need to get to work."

# 21

# INTERNAL QUALITY REVIEW

## 21.1  MEETING WITH GAIL STROM AND MARCIA HINES

The team gathered the required information from the implicated inhaler lots and found that the incoming inspection of the container and meter dimensions showed they were within acceptance tolerances; though the meter orifice was on the low side, it was still acceptable. Bruce pulled samples from the five lots manufactured to date and the acceptance records met the final product release label claim, though the actual testing records indicated the dosage testing in each lot to be 10 percent over the manufacturing tolerance specification. The team included additional exploratory testing to verify the dosage delivered over the total applications contained in ten randomly selected inhalers from the implicated lot. The exploratory testing confirmed that the dosages were on average 10 percent over manufacturing acceptance with a maximum of 12 percent but still the average was within label claim. On further review of all manufacturing documentation it was discovered that the first lot to stock was actually an aliquot from the 50-milligram reagent lot and not the 25-milligram lot as thought. The filling documentation was written for the 25-milligram lot with the associated 25-milligram labels.

*A History of a cGMP Medical Event Investigation*, First Edition. Michael A. Brown.
© 2013 John Wiley & Sons, Inc. Published 2013 by John Wiley & Sons, Inc.

The data was compiled by the team in six days and the potential causes that could contribute to a defect in the inhaler were noted in the contradiction matrix.

"Gail, the team has completed the internal investigation following our MAIC procedure. We weren't able to identify a defect in the inhaler device itself but there are factors that could contribute to the death implied in the Agency Medical Event Report."

"I figured as much, go on."

"Just to summarize our findings, I'll go over the potential causes with the facts that contradict the cause due to a defective inhaler . . . or in two cases, could actually be the cause.

"Overfilling of the container was contradicted by lot release volume testing. The acceptance records for the inhalers from the implicated lot were within final product fill volume ratings. We did additional testing of ten inhalers with identical results. This eliminates this cause.

"Dimensions of the metering section: The incoming quality testing for each component in the metering subassembly was within acceptance criteria. This eliminates this cause.

"Fit of the metering section with the container pump mechanism: Again, the acceptance criteria were met, as documented in the final lot acceptance. All ten samples from the implicated lot were disassembled and no interferences were detected. This eliminates this cause.

"Increased dosage over label claim: We have two serious issues here. First, we discovered the implicated lot was mislabeled: the inhalers were filled with a 50-milligram aliquot but the labels were for a 25-milligram dosage. Second, the manufacturing records show that on average the dosage of all three released lots was 10 percent higher than the acceptance criterion but still within label claim. Though the implicated lot was labeled at 25 milligrams the testing should have shown a much higher dosage variation. This was verified by additional testing using the ten randomly selected inhalers from the implicated lot and does not make sense. This testing data should have shown a significantly higher variation and the lot should have been put on quality hold immediately based on the label, but the release testing was based on a 50-milligram fill. The Agency will note these as violations of cGMP* requirements . . . label and internal reagent lot manufacturing control are major cGMP violations."

Marcia hesitated, "And, based on these violations, just the filling volume variance alone, all five lots should have been placed on quality

*cGMP = current good manufacturing practice.

hold for further investigation. I strongly suggest that all lots be put on quality hold for label and lot reinspection.

"The modified Katlin inhaler may not have been tested adequately, as pointed out by Engineering, but all qualification data and the NDA* clinicals were reviewed and the inhaler passed all acceptance criteria. There were no aberrant patient results identified by the obstetrical–gynecological . . . OB-GYN . . . investigators. This eliminates this cause. The design team meeting documents are weak. The minutes indicate neither the Quality nor the Engineering organization was in attendance for the key design decisions: another cGMP violation.

"Improper use of the inhaler: On examination of the product insert, there are no limitations on inhaler use. Medical Affairs confirmed the clinical human populations were not segregated for lifestyle, race, etcetera, or for potential interaction with other disease states or medications. The clinical focus was on patients who showed signs of child abandonment and irritability. This puts the NDA at risk and may raise questions on the validity of our product insert. Medical Affairs is contacting each of the OB-Gyno clinical investigators. The Agency could require the product to be taken off the market until a new insert can be developed. We may want to include an implementation timeline in the investigation to correct the insert.

"Medical Affairs brought to the team's attention that the foxepin toxicity studies were done with oral administrations, not injections as we submitted in our NDA. The contradiction matrix eliminates this as a cause based on the human studies in our NDA. I personally don't feel this is adequate. This may trigger an Agency audit by itself."

Gail let Marcia go over the complete investigation without comment until she was done. Gail asked, "What about the fact that the inhaler was designed, manufactured, and tested and the NDA completed in less than two years. Didn't your team think this was suspect and should be part of this investigation?"

"As a matter of fact, it was brought up but the team decided that it wasn't pertinent to a quality issue. Keith Carlisle was able to convince us all it was just a good example of concurrent design and accepted in the industry. If it was an issue, the qualifications wouldn't have been successful."

Gail said one more thing, "Set up an executive management review."

*NDA = New Drug Application.

## 21.2 EXECUTIVE MANAGEMENT REVIEW

### In Attendance

M.V. Brooks, Kinnen Chief Executive Officer

R.L. Siegal, Pharma President

Dr. Tom Watson, Pharma VP R&D

Dr. Susan Vickory, Pharma VP Medical Affairs

Larry Fletcher, Pharma VP Operations

Gail Strom PhD, Pharma VP Quality and Compliance

Marcia Hines, Pharma Quality Director

Marcia opened the meeting with a short agenda covering the seven potential causes and the contradictions that eliminated the inhaler design as the cause, with a focus on the cGMP violations, the mislabeling, the mixing up of the fill lots in manufacturing, the potentially incomplete patient information in the product insert, and the university toxicity studies.

Brooks asked, "Is there any way that our internal violations can be hidden?"

Marcia had the floor but looked at Gail Strom to answer. Gail answered, "Mark, the investigation has been reviewed at the Pharma management level and signed off. The cGMP violations by themselves will trigger an audit. When the Agency is in here they'll review the acceptance records and note the mislabeling, the screwup in filling the wrong lot definitely will be found, the toxicity information from the university is bound to come up, and they'll do a thorough review of the product insert.

"From a quality perspective I recommend all finished-goods lots be placed on quality hold immediately and we do a voluntarily recall of all product in the field."

Brooks thought about it for a moment. "No. The product is selling better than expected. We need to maintain this pipeline. If we put the product on quality hold or recall, it will lead the Agency right to the heart of the problem."

Gail hid her disappointment in Brook's lack of quality concern—the product should at least be put on hold—and continued. "We may as well, in a show of good faith, present the information as part of our internal investigation. If we don't, we risk an Agency warning letter. I'll bet the Agency already has the university study. Misrepresentation of data to the Agency investigation team could not only lead to a warning letter but also put the entire division at risk of a consent decree. A consent decree could cost us millions of dollars as well as put us under Agency investigation scrutiny for a minimum five-year period."

Larry Fletcher asked if he could speak; it was unusual for Larry to ask but he did. The anger in his eyes was obvious. "Gail, I didn't review any of this investigation and as Pharma Operational VP this is not acceptable."

Gail looked Larry directly in the eye, squinting somewhat for effect and said quite calmly, "Larry, the investigation is under the direction of Quality and Compliance, the only required Division review is at the Director level of the investigation team . . . It is our policy. Each Director, including Keith Carlisle, had a chance to review. The only signatures required are mine, Marcia's, and Garvey's as technical lead. If you have a problem with the results of the investigation, talk to Keith."

Gail maintained eye contact with Larry until he looked away. In her opinion Larry was the biggest problem that Pharma had.

The meeting lasted twenty minutes. Brooks needed to start damage control at the Agency's D.C. office.

\*\*\*

Brooks' admin let him know that she was able to reach Dr. Koppel and he'd return the call at four that afternoon.

Brooks was in his office ten minutes early with Dr. Vickory, waiting for the Agency Deputy Commissioner to call. While the two waited, Dr. Vickory informed Brooks that Gonzales had met with the clinical investigators to review the human studies and would begin updating the patient insert, most likely based on theoretical interactions of a typical tricyclic.

The call came in promptly at 4 PM.

"Louis, haven't spoke with you for some time. We need to get together soon for golf at my club. Bring the wife and kids . . . they can spend the day at the pool. I'll have the company plane pick you all up in D.C. By the way, you're on speakerphone and Dr. Vickory is in my office."

"Susan, how are you . . . Mark, a lot on my plate these days. I could possibly get away for a long weekend at the end of the month. We should spend some time together. I'm guessing this call is about the letter requesting an investigation on the Baltimore Medical Event."

"Unfortunately, you're correct. We've completed the investigation and need your advice on how to present the data. I'm embarrassed to have to say that the team uncovered a few cGMP violations and a possible issue with the NDA submittal and product insert."

Dr. Koppel was silent for a moment, finally asking, "How bad is it?"

Brooks answered, "I need you to look it over and tell me how the Agency will react. Personally, in my opinion, the findings are minor but, if not handled by your D.C. office, could get out of hand."

"Send me a copy marked 'Confidential—For Review Only.' Have this in my office by end of day tomorrow. Give me a day or two and I'll get back."

Brooks said, "Thanks."

\*\*\*

Dr. Koppel sat in his office thinking. *He needed to review the investigation . . . The IND\* and NDA were approved by his office and he personally approved the abbreviated clinicals . . . if there're, issues he needed to know immediately. The Medical Event was a death implicating a Kinnen product. The Baltimore office received the complaint and he was certain they would recommend a cGMP audit no matter what the Kinnen internal investigation revealed. If the issue with the NDA and product insert were serious, it would be handled through Regulatory Affairs. If he interfered, it wouldn't look good. An election was one year off and the Agency was under Senate scrutiny as it was. Whichever party takes over the presidency and Senate there could be a new Commissioner appointed. He needed to keep himself as far away from this situation as possible. And there is no way he will meet with Brooks on the golf course.*

\*\*\*

Carole answered Brooks' phone two days after the initial conversation with the Agency Deputy Commissioner. Dr. Koppel's admin was on the line asking to set up a call later that day.

Brooks, Siegal, and Dr. Vickory participated in the call with Dr. Koppel. The call was short and to the point.

"Louis . . . Sue and our Pharma President, Bob Siegal, are on the line as well. What's your take on our internal investigation?"

Dr. Koppel answered, "Your investigation is in order . . . but there are concerns. I'm glad you approached this honestly, bringing each potential issue out in the open. This can only help. The issues are serious. I suggest you submit your investigation exactly as prepared."

Brooks did not like to hear this and asked, "How serious?"

"The issues are serious enough to warrant a complete cGMP audit as well as a review of the NDA. It could force you to take the inhaler

\*IND = Investigational New Drug.

off the market and recall all product in the field. I'm especially concerned with the mislabeling and mixup between the two dosages. The absence of the toxicity study even if there were no patient interactions in the clinical studies may not be an issue but will be addressed. The patient insert can be dealt with."

"Louis, can you intervene?"

Louis did not answer. Brooks asked, "Louis, not sure if you heard me."

Louis answered, "Mark, I heard you. This is out of my hands. The Agency's Regulatory Group and the Baltimore Regional Office will handle the investigation. Baltimore may handle Regulatory as well. I cannot intervene. My office has helped you in past audits but this involves a death. This is public record and has to be handled to the letter of the law. If not, we could all go to jail."

"So you're sure there will be a formal investigation."

Dr. Koppel did not hesitate in his answer, "Exactly."

# 22

# THE AGENCY AUDIT LETTER

The investigation was submitted to the Baltimore office, leaving out the violations of current good manufacturing practice (cGMP). A registered letter from the Agency was received one week following submittal.

The letter explained that, after review, the Baltimore Regulatory Office would be on the Kinnen campus within the next two weeks to conduct an audit that could last up to three months. No specific arrival date was given.

The letter instructed Kinnen to have records regarding the Oxy-Fox Inhaler design and equipment qualifications and the finished-goods quality inspection reports for all manufactured lots available for their review, and copies of all launch and product development team meetings. There was no need to have copies of the New Drug Application (NDA) available as it is currently under review by the Agency. In addition, Kinnen must make available to the Agency all external customer complaints and all internal manufacturing nonconformance documentation for the entire Pharma product line.

The letter identified the investigation team, with Dr. William Slone, Regional Director Baltimore Office, as the Principal Investigator. The investigation team includes experts in equipment qualifications, product

*A History of a cGMP Medical Event Investigation*, First Edition. Michael A. Brown.
© 2013 John Wiley & Sons, Inc. Published 2013 by John Wiley & Sons, Inc.

validation and quality control, pharmacology, and cGMP regulations—all from the Baltimore office.

\*\*\*

After reading the Agency letter, Brooks called Siegal, Sue Vickory, Larry Fletcher, and Gail Strom to his office. Copies of the letter were given to each one.

Brooks began, "Gail, you will be our contact with the Agency during the investigation. You need to have all the documentation available as requested. I'm concerned that the request includes all past Pharma complaints and quality records related to nonconforming products identified in manufacturing. Any comment?"

"At your direction, we took out specific reference to cGMP issues but the investigation implied cGMP issues. There was no way to really hide them. In this type audit it is normal for the Agency to look at records for our complete product line. The Agency wants to determine if this is systemic to our entire operation or just a single compliance failure."

Brooks continued, "I guess that's why the Agency expects to be on our site for three months. Siegal . . . you and Gail identify the team that will interface with the Agency investigators. Make the team small . . . no more than one person from each area should have face-to-face contact. I want a rapport built between us and the investigators. Set up a command center and block out at least four conference rooms for the entire three months. Make sure you have breakfast and lunch brought in each day. Make them comfortable . . . make them our friends. The command center must be able to bring up any electronic document from our files as the Agency requires.

"Assign people from the quality group to escort the investigators . . . no way let them wander around our site unescorted. Also have people from your respective organizations available in the command room. Any request from the Agency must be in writing and notify the key people involved that they will be here for the duration . . . all time off, including vacations, is on hold. And in no case shall an Agency request take more than two hours to get supporting documentation . . . Gail, what's your plan?"

"I've set up meetings beginning this afternoon with the entire quality organization. Meetings with the product development people and a number of the engineers and production people who may be involved will begin tomorrow. Marcia and I will handle these meetings. The main focus will be how to answer Agency questions. In the past when a question is answered, the Agency investigator will not respond but just make eye contact. This is a ploy to make the person nervous . . . continuing to talk . . . bringing issues into light that have no bearing on the question, offering the Agency a lead into another area. The central point in these meetings is to instruct those being questioned to answer the question directly and then shut up . . . no elaboration unless requested to do so. We'll also have our experts available to counsel the face-to-face Agency interface people prior to any meeting and actually sit outside the room during the meeting. These face-to-face people will be advised to ask for a short recess to consult with the experts if they cannot answer a question.

"My Quality group will meet first. I'll have a list from each Pharma VP and Director by tomorrow morning and these people will go through the training I mentioned."

Brooks continued, "I want a Pharma management review at the end of each day the Agency is on-site. Have a list of all Agency questions, any documents requested, who in our organizations took the Agency questions, and the response. Try and get a feel for the Agency's position as well. Siegal, you will have a report in my hands at the beginning of each day summarizing the Pharma management meeting. Questions?"

There were no questions. Larry Fletcher kept his mouth shut. *He knew that the cGMP violations were in his camp and he'd eventually be held responsible. Product availability was his primary concern and he had personally directed the first lot to stock not to be put on hold but issued to finished-goods inventory . . . unfortunately, this was the implicated lot . . . if the product went on hold, the launch schedule would have been delayed . . . he could not allow any delay . . . Give me a break, how bad could a 10 percent overfill be? The mislabeling was new to him and the mixup in filled lots could be a real problem. The lab testing should have discovered that the lot was actually a 50-milligram dosage from the filling documents instead of the labeled 25 milligrams. This mixup is not his fault. He was also concerned that the Oxy-Fox meeting minutes would be thoroughly reviewed by the Agency and this could lead to a number of issues that may prove to be embarrassing.*

# 23

## AGENCY ARRIVAL

Dr. William Slone arrived on the Kinnen campus precisely at 8 AM on Monday morning of the second week following the Agency letter. Accompanying him were subject matter experts from the Baltimore Regulatory Office. There were ten investigators.

Dr. Slone displayed his credentials to the guard at the visitors' entry gate. The guard had been told to expect the Agency at any time and to inform Gail Strom on their arrival.

Gail went to the visitors' gate herself to greet the Agency investigators. She was somewhat shocked to see that the entourage included ten inspectors; she had only accounted for six based on the letter. The guard took an inordinate amount of time to prepare the identification passes that would allow the Agency entry to the Kinnen premises. Gail thought to herself, "*This is not a good way to begin.*"

She greeted each person respectfully and asked them to follow her car to the main operations building, where the meeting would be held. When the group entered the operations building, Gail stopped them in the main entry area.

"Before we go any further, I just want to welcome you all to Kinnen. I've set up conference rooms you can use during your time with us and have my entire organization on call and at your disposal. The first

*A History of a cGMP Medical Event Investigation*, First Edition. Michael A. Brown.
© 2013 John Wiley & Sons, Inc. Published 2013 by John Wiley & Sons, Inc.

thing we'll do is meet in the executive conference room for introductions and I want to propose an agenda we could follow over the next few months."

Dr. Slone answered, "Gail . . . thanks. Hopefully we won't need to be here that long, but one never knows. Let's get started."

Gail led the group to the conference room on the main floor. The room is designed to hold up to thirty people at full capacity so there's more than adequate space. The room itself is constructed of heavy mahogany paneling and there is an oval table capable of seating fourteen people, with additional chairs located around the outer walls. The chairs are soft leather, designed for comfort as the meetings usually conducted in this room could last for many hours. There's a service bar in the corner and a whiteboard hidden behind a panel with swing-out matching mahogany doors. A projector hangs above the table and a screen concealed in the ceiling could be automatically brought down for viewing. The room is totally soundproof.

Several people were in the room waiting: M.V. Brooks, Kinnen Chief Executive Officer; R.L. Siegal, Pharma President; Dr. Tom Watson, Pharma VP of Research and Development; Dr. Susan Vickory, Pharma VP of Medical Affairs; Larry Fletcher, Pharma VP of Operations; and Marcia Hines, Pharma Director of Quality and Compliance.

Gail told the Agency people to make themselves comfortable and indulge in the goodies on the service bar. Dr. Slone acknowledged but said, "No, thank you. This is an official audit and the only refreshment we will be interested in is water."

Once introductions were completed Gail displayed an agenda on the overhead screen and began the presentation.

"To begin, I'm Dr. Gail Strom, Corporate Vice President of Quality and Compliance for the Pharma Division. The agenda presented is only a suggestion. Our organization is prepared to make any deviations the Agency requires. First we suggest we review the internal investigation leading to this audit. Second, we have the complete files for the inhaler qualifications and production lot quality acceptance documentation that you've requested. Once they're reviewed we can go over the complete Kinnen Pharma customer complaint history for the last two years. I'm sure you're aware that the local Agency office was in here two years ago and reviewed the complaints prior to that time. After you've completed this review, our internal nonconforming-goods documentation will be made available. I suggest we conduct some of the audit in parallel. We have four conference rooms available to do this . . . none as nice as this room but they are equipped with the necessary communication systems that will be needed. Any questions or concerns?"

Dr. Slone answered, "For the time being your agenda is fine. At this point, I need to advise you everything said in these meetings will be recorded for the public record and could be used in a court of law."

Gail held her breath for a moment. She then said, "Understood."

Dr. Slone went on, "We'll begin meetings tomorrow. My group needs to check into the hotel this afternoon. We'll be here at 9 AM. As you suggested, the teams can split up and review the documentation concurrently. I want the Agency Qualification Specialists to review the complete Oxy-Fox Qualification documents, including the Master Plan, the Installation Qualification, Operational Qualification, and Process Qualification. The cGMP experts will review the lot information for the inhalers issued to the clinical investigators, first-lot-to-stock quality acceptance records for the market release inhalers as well as the same records for the implicated Medical Event lot if not the same, and the acceptance records for all subsequent lots. A third group will review all team minutes leading to your inhaler launch, including the design and program team meetings and the management reviews. I want to review the Katlin due-diligence report myself . . . we've seen the IND[*] and the NDA and are very concerned with the absence of toxicity studies in the animal studies, and I want to see the Design History file—including the Verification and Validation protocols and results. I suggest you marshal your experts and have them ready to answer any of our concerns.

"As you know, when we need an answer, the question will be given to you in writing and we'll allow you no more than two hours to research the question. If you need documentation to assist in your answer, it must be the original signed document. We will not accept copies.

"The meetings will last no more than four hours with bio-breaks, and of course, time for lunch. We'll eat in your cafeteria by ourselves. I don't want any of your people with us and I want an area cordoned off for privacy. At the end of our day, at approximately 3 PM, the lead Agency investigator from each meeting, as well as myself, will meet with you and no more than two of your quality or technical people to review our findings from the meetings. One hour will be allowed for this meeting. Any observations we discover will be brought to your attention and if corrective actions are taken with appropriate documentation prior to the end of this audit, the observation will be identified in our final report and noted as corrected. Do you understand?"

Gail answered, "Clearly."

---

[*]IND = Investigational New Drug; NDA = New Drug Application.

The Agency people excused themselves and were escorted to the main gate by Susan Jaffey.

\*\*\*

Brooks asked the Kinnen group to stay a while.

Brooks asked, "Gail, your take on this meeting?"

"Mark, it's very unusual to have a Regional Director as the lead investigator. And the subject matter experts are all from the Baltimore office, which received the Medical Event complaint. Dr. Slone is no novice in this role and his medical background is in cardiology. I'm more than concerned that he's reviewing the product development documentation himself. He'll definitely key in on the Katlin study for which Kinnen has no real defense. Our due-diligence team accepted this report at face value without reviewing the university studies for the foxepin. It was assumed that foxepin was a typical tricyclic and completely characterized for intramuscular injection. Dr. Slone has the IND and NDA and will more than likely go through the human study data with a fine-toothed comb. We could be in trouble."

Sue Vickory interrupted, "I agree with you somewhat. But the NDA did test patient populations with no adverse affects and we did present authoritative literature supporting the use of tricyclics. I reviewed it with the D.C. office, and the Agency approved abbreviated clinical studies based on the literature. My take is that in the worst case we'll have to remove the Oxy-Fox from the market until an updated product insert is approved indicating potential usage limitations and interactions . . . And Gonzales is working with the OB-Gyno investigators as we speak. This will definitely be identified as an observation but we should be able to rewrite the insert before the Agency leaves, and as Dr. Slone pointed out, it would be noted as a corrected observation with no additional action required."

Gail answered, "Sue, cardiology is your area. You need to be available to meet with Slone . . . cardiologist to cardiologist."

Brooks asked, "Any other opinions?"

Gail answered, "Even if we get through the NDA and insert issues, the cGMP violations will definitely result in a warning letter."

Brooks answered, "We took the cGMP issues out of our internal investigation report. It's a roll of the dice if they find them or not. If this is all we have to worry about, we can handle a warning letter. Letters are common in this industry and are of no real concern . . . at least not to me. Unless the Agency makes a big deal publicly this will have no impact on our company."

# 24

# THE AUDIT

## 24.1 AGENCY MEETING TO REVIEW QUALIFICATION DOCUMENTS AND THE QUALITY ACCEPTANCE RECORDS OF FIRST LOT TO STOCK

The first meeting was to review the qualification documents and to review the quality acceptance records of the first lot to stock.

### In Attendance

Ronald Symms, Lead Agency expert on Current Good Manufacturing Practice (cGMP)

Dominic Papas, Agency cGMP expert

Paul Fisher, Lead Project Engineer for Oxy-Fox equipment

Bruce Garlin, Kinnen Site Plant Manager

Ronald Symms began, "Dominic and I reviewed the Master Plan and the supporting qualifications. As you know cGMP regulations require multiple lots to be tested in the Operational Qualification and Process Qualification. We can only find two lots in each. Can you explain this?"

*A History of a cGMP Medical Event Investigation*, First Edition. Michael A. Brown.
© 2013 John Wiley & Sons, Inc. Published 2013 by John Wiley & Sons, Inc.

Paul Fisher answered, "I think you may be looking at the traceability of the raw material lots. Actually both material lots were used in each qualification with an additional run with the second material lot. There are three lots in total for each qualification. If you look in the section for each qualification you will see that each run is identified."

Ronald Symms looked at both qualifications and agreed.

Paul asked, "Does this answer your question?"

"For now, thanks. We also noted in the finished-goods quality acceptance record the implicated lot that initiated this investigation was on average 10 percent higher than the manufacturing tolerance, with a high of 12 percent, and was not put on quality hold. Please explain."

Bruce took this question, though with considerable hesitation. "Even though the lot was outside the acceptance range for production the product was within final product release specifications. We have the option to write a deviation and, if approved by the Quality group, the lot can go into finished-goods inventory. This deviation is explained in our plant Quality Procedure."

Symms continued, "Please provide your plant Quality Procedure to us and the signed deviation. I'm not able to find the deviation in the lot history file."

Symms wrote down the request and gave it to the Kinnen administrative assistant to pass on to the command center.

Symms watched her leave, thinking, *"It is good to be in the field"* . . . *he loved these investigations and has come up with a potential observation.*

Bruce asked, "Is that it?"

Symms answered, "Not quite. When we reviewed the acceptance records for the implicated lot, the lab testing was at a 50-milligram dosage . . . label copy attached to the lot indicated a 25-milligram dosage but when we traced the documents in the production record, we discovered that the lot was a 50-milligram aliquot. Please explain."

Bruce had prayed the filling mixup wouldn't be discovered. He answered truthfully, "I have no explanation. Our label control system was validated and 'bulletproof'. Label control is a double-approval system. When labels are issued to the production line, the operator checks the production documentation for the fill and dosage, matches the label to the documentation, and obtains supervisor review and signature that the label information matches the production fill. I can't explain how this could happen. An identical procedure is required for each aliquot given to the filling line."

Symms didn't say a word, just continued to make eye contact with Bruce. Bruce kept his mouth shut and looked at the floor. What else could he say!

Symms answered, "Your explanation on label control is good and follows cGMP but you did not address the filling lot. You need to get this documentation and explain how your internal control system failure occurred. I also want the quality acceptance records for all inhaler finished-goods lots produced from market entry until now. Dominic, do you have anything to add?"

Dominic did not.

\*\*\*

As Bruce left the room, he knew that there was no approved deviation and the mixup in filling lots was inexcusable. *And how could the testing lab have missed the dosage? He wondered if he could have answered the questions better . . . possibly not mentioning that the deviation required a quality approval.*

## 24.2   AGENCY MEETING TO REVIEW THE OXY-FOX INHALER LOT USED IN THE NDA CLINICAL STUDIES

### In Attendance

Ronald Symms, Lead Agency cGMP expert

Dr. Brigitte Holander, Agency Medical Pharmacologist

Dr. Jeremy Sawyer, Agency Regulatory Affairs

Dr. Melissa Green, Agency New Drug Application (NDA) Consultant

Dr. Susan Vickory, Kinnen Medical Affairs

Dr. Jim Gonzales, Kinnen Medical Affairs

Ronald Symms opened the meeting. "We've reviewed the NDA submission and the Validation Master Plan for the Oxy-Fox Inhaler. Specific questions were given to your admin last night and I assume, since you are here, you can provide an explanation. Our first question concerns the manufacturing dating on the inhalers delivered to the OB-Gyno investigators. The dates do not align with the completed qualification dates in your Master Plan. Please explain."

Jim Gonzales answered. "We followed a parallel approach so that the inhaler qualifications were done concurrently with the inhalers used in the clinical studies. I have copies of the final prototype qualification documents as well as the final production records that verify the inhalers were produced on a validated prototype system and that the prototype system was equivalent to the final manufacturing process."

Dr. Sawyer from the Agency Regulatory Affairs Office asked Gonzales, "Dr. Gonzales, in your position in Medical Affairs I assume you're aware that all products used in human studies must be validated prior to commencing these studies. Are you aware of this Agency requirement?"

Gonzales answered, "I'm aware of the requirement. The program launch team made a decision to reduce the launch timing. The processes are simple and I agreed that the risk was minimal."

Dr. Sawyer continued, "We have copies of team meeting minutes, specifically Team Meeting Number 3. Please refer to the discussion item from Sue Jaffey, the Quality group team member, which I'll read . . . 'The Agency requires complete validations on all lots issued for clinical studies'."

"From the content of the minutes, Ms. Jaffey is referring to the clinical lots given to the NDA investigation sites. Is this correct?"

Gonzales answered, "Yes that is correct."

Dr. Sawyer continued, "So the product given to the NDA investigators had not been through the completed validation process and there was no objective evidence that the product was safe or effective at the time of release. Is this true?"

"Well, if you take it literally. That is true as well."

"Dr. Gonzales, I assure you the Agency takes this literally."

Ronald Symms continued the audit, "Actually that answers our second question as well. This meeting is finished.

## 24.3   AGENCY MEETING TO REVIEW THE DESIGN AND PROGRAM TEAM MEETING MINUTES

### In Attendance

Jerry Ramsey, Lead Agency cGMP expert
Donald Orloff, Agency cGMP expert
Dan Garvey, Director Product Development
Keith Carlisle, Director Program Management
Susan Jaffey, Quality

Jerry Ramsey started the meeting. "We've reviewed the meeting minutes for the design of the inhaler. Our questions were given to your admin early this morning. You have these in front of you now. I want an explanation of why the engineering and quality functions were not in attendance at the majority of these meetings.

I specifically asked to have an Engineering representative in this meeting."

Garvey had given considerable thought to how he should approach this inquiry and reviewed his answer with the Kinnen cGMP expert assigned to the group. "I was the Engineering rep on the market launch. My position as Director Product Development wasn't until after market launch. I've given considerable thought to this question and will answer to the best of my ability. Engineering's function is to assure that the inhaler can be manufactured. It is not necessary to be in attendance for each design meeting."

Ramsey then asked, "Being the Engineering rep on the design team I would expect you to sign off on the final design. The signature is the Engineering Director. Can you explain?"

Garvey had thought about this question before the meeting and, even with counsel from the expert, neither could come up with a good answer. Dan decided to use the manufacturability argument. "I was not happy with the final design. I thought there were too many components in the metering section and did not sign off from a manufacturability concern."

Though Ramsey was an Agency cGMP expert, his background was in medical device design, and he asked the next question. "The final inhaler design is a modification of an on-market device. The meeting minutes, actually Meeting Number 7, written by Gordon Taylor, indicate you expressed concern with the design and suggested that the inhaler be completely tested for nasal administration at the prescribed dosages. Please explain your position?"

At this point Keith Carlisle interrupted, though he should have kept his mouth shut and let Garvey answer. "My position is currently Director of Program Management. I was the Program Manager for the development and launch of the inhaler. You are asking Dan his opinion. His opinion is not pertinent to the design of the inhaler. It was not his job at the time."

Ramsey was quiet for over a minute just making eye contact with Keith. And as Ramsey expected, Keith continued to talk. "The inhaler was fully qualified and meets all product release specifications. That is all you need to know."

Ramsey said, "Okay."

Ramsey made a note to get Keith Carlisle into the meetings with Dr. Slone, thinking, "*Carlisle has an attitude we can take advantage of. Slone knows how to push his type's buttons.*"

Ramsey continued, "Keith, we've reviewed the program team minutes and find them in order. I have two questions. The first is about

your decision to follow a concurrent design of the inhaler using prototype plastic parts and prototype processes for the NDA required clinical studies. This practice is common with medical devices but not with
pharmaceuticals. The Agency expects all compounds used in NDA
clinicals to be manufactured on final, validated processes . . . not prototype equipment that has not been validated. And second, the minutes
repeatedly request Engineering to reduce the development time. My
call is this is no way to run a program."

Keith started to get agitated. Ramsey had worded the question to
see how far Keith could be pushed. Keith answered, "Our design procedures allow the teams to follow a parallel path. The prototype systems
were qualified under the same documentation as the final production
equipment. If there is an Agency regulation specifically stating this as
a violation, show it to me. Secondly, you have obviously never run a
market launch program. My job is to get the work done within schedule
and budget. Do you understand?"

"Yes, I understand. I also understand from these minutes that you
were trying to intimidate Engineering to take shortcuts in the process
design."

Ramsey also knew how to push buttons.

Keith was visibly upset and said, "All you need to know is that our
product qualifications met all acceptance requirements. It is none of
the Agency's business how we run our programs."

Ramsey responded, "That's all we need from you now. I'm writing a
formal request to review your internal Quality Policy in regard to
product development. Thanks for your time . . . and Keith . . . we'll see
you later."

\*\*\*

Garvey and Carlisle left the meeting. The two walked down the hallway
leading to Dan's office. Dan said, "Keith, you just went on public record
as a complete jerk. You, by yourself, will blow this whole audit."

## 24.4   AGENCY MEETING TO REVIEW THE DUE-
## DILIGENCE REPORT, KATLIN STUDIES, AND OXY-FOX
## DESIGN HISTORY FILE

### In Attendance

Dr. William Slone, Lead Agency Investigator
Dr. Brigitte Holander, Agency Pharmacologist

Dr. Jeremy Sawyer, Agency Regulatory Affairs

Dr. Melissa Green, Agency NDA Consultant

Dr. Susan Vickory, Kinnen Pharma Medical Affairs

Dr. Jim Gonzales, Kinnen Pharma Medical Affairs

Dr. Gail Strom, Kinnen Pharma Quality and Compliance

Marcia Hines, Kinnen Pharma Quality

Keith Carlisle, Kinnen Pharma Program Management

Patty Keyser, Kinnen Pharma Marketing

Following introduction of the meeting participants, Dr. Slone started.

"Our team has reviewed the due-diligence report, and the IND[*] and NDA documents. There are a number of issues. Your Design History file is in order but the due-diligence report is questionable.

"This is the third meeting on these topics. We met with Jeff Daniels in your R&D group, Dave Stall in Materials Management, and Gordon Taylor from Product Development. To be perfectly frank, the meetings raised a number of Agency concerns.

"You've had our team's specific questions in your possession for more than two days without an answer. This in itself will lead to an observation.

"I want to begin with the due-diligence report. I've specific questions for Keith Carlisle. I want to remind you all that this meeting is being recorded.

"Keith, from testimony of the members of your team I gathered you were in charge and made the decision to accept the Katlin data even though some of the data was questionable. Also the team failed to review the university toxicity report on the foxepin. Can you explain this?"

Keith had never been in a situation such as this and he was visibly nervous. He began defiantly, "Dr. Slone, this is not a court of law and I don't understand your use of the term testimony."

Dr. Slone answered, "Your organization should have prepared you. Everything said in these meetings is on the record and can be used in a court of law if necessary. This is definitely testimony and has been recorded as such. In the three previous meetings, as well as this one, the young man sitting in the corner of the room is a licensed court reporter. Now please continue and answer."

Keith's buttons were being pushed.

---

[*]IND = Investigational New Drug.

"The R&D group was actually the lead in the due-diligence report. Jeff Daniels was the lead. My role was to assure that the acceptance schedule was met."

Dr. Slone continued, "Did you influence the acceptance of the data in order to meet the schedule?"

"That's my job. The team didn't have the experience in animal studies and were confused in what they should be looking for. It is my job to manage these activities and in the confusion I stepped up and took the lead."

Dr. Slone asked, "Is your background in a scientific discipline so that you're qualified to make decisions of this type?"

"I'm an engineer by training and, I think, well qualified to make those decisions. I mean, how difficult is it? The mouse study determined the Oxy-Fox cocktail to be effective in eliminating postpartum depression. That was the object of our review . . . simply that. I don't understand why you're making a big deal."

Dr. Slone answered, "A young woman with three children may have died from an overdose of the Kinnen Oxy-Fox Inhaler. That is why I'm making a big deal."

Dr. Slone held his gaze on Keith for a few moments, and then said, "To sum up . . . you have no scientific background . . . you used your position as Program Manager to influence the team to accept the mouse study without any further evaluation of whether toxicity had been established or multiple species studied . . . and you did this because you needed to meet the schedule. Is that a correct statement?"

Keith answered, "Okay. I took charge. Someone had to."

Dr. Slone asked the last question for the meeting, "I asked for a representative from your marketing organization to attend this meeting. I personally reviewed the market launch plan and am concerned with the details on how 'free samples' would be given to the OB-Gyno physicians . . . specifically the reference to the educational series. I want to see your sales records indicating the names of the physicians who received these samples."

Patty had reviewed the Agency question with the consultants prior to the meeting and was assured this was not an area subject to Agency audit. She was advised to give a brief review only. Patty answered, "Dr. Slone, my name is Patty Keyser. I'm handling the marketing and sales for this product. I reviewed your written request and don't quite understand the bearing on this audit. Can you clarify your request?"

"Patty, we need to know the names of the physicians who received samples of the Kinnen inhaler. They may have to be notified that the drug is under Agency investigation and potentially unsafe."

Patty answered, "I'll have the physicians' names to you within the hour. The educational series describes how the Oxy-Fox Inhaler is to be prescribed ... the medical effect of both dosages ... with specific reference to patients who demonstrate irritability and tend to ignore the child."

Dr. Slone said, "That's fine. All I require is a copy of the series and the physicians' names."

Gail Strom asked, "Dr. Slone, do you have additional questions?"

"No. We're done for the day. I want an hour to review our notes and meet with you before my team leaves."

\*\*\*

As soon as the meeting was over, Patty called Maria Sanchez, Pharma Director of Marketing and Sales. Maria was in a meeting; Patty told her admin it was an emergency and she had to talk to her immediately.

The admin located Maria and she took the call.

"Maria, this is Patty. I just came from a meeting with the Agency. They've requested the names of the physicians we gave free samples of Oxy-Fox, and the series telling them how to prescribe the drug. You know that the series includes specific instructions on how to obtain insurance company reimbursements for the free samples. I'm sure it will be considered an ethical violation."

Maria answers, "Patty, we'll let Siegal handle it. The entire physician reimbursement plan is his idea. Thanks for the call ... I'll pass this on to Susan DeAngle" (Pharma VP of Marketing and Sales).

# 25

## END-OF-DAY AGENCY WRAP-UP MEETING

**In Attendance**

Dr.William Slone and Dr. Jeremy Sawyer, Agency

Dr.Susan Vickory and Dr. Gail Strom, Kinnen

Dr. Slone began, "I'll make this short. My team will take a week's break starting tomorrow.

"Up to this point the audit has determined that your finished-goods acceptance practice and lot control are in violation of Current Good Manufacturing Practice (cGMP) regulations. Based on our review of the Oxy-Fox lot implicated in the Medical Event, Kinnen moved a nonconforming lot to finished goods without additional testing or quality review and sign-off. Review of additional inhaler finished-goods acceptance records indicates this practice is systemic. And of major concern is that your lot control and accountability procedures are not being followed in production. These are both audit observations.

"The Medical Event Report indicated that the patient was prescribed a 25-milligram dosage. The Agency has completed our own testing of the inhaler implicated in the death, confirming that the dosage administered in a single application was 50 milligrams . . . The

*A History of a cGMP Medical Event Investigation*, First Edition. Michael A. Brown.
© 2013 John Wiley & Sons, Inc. Published 2013 by John Wiley & Sons, Inc.

implicated product was mislabeled. The patient was reported to have taken a double dose. At 25 milligrams, the second application may not have had an affect . . . but at 50 milligrams, a second inhalation into the nasal passage may have contributed to cardiac failure. As scientists and medical professionals you should examine your obligations to patient welfare.

"Acceptance of the Katlin due-diligence report was highly prejudiced by the Program Manager. The Program Manager is inadequately trained to serve in this role; cGMP regulations require your people to have the appropriate training. This is an audit observation.

"Toxicity studies were not done for an injected foxepin compound. This is not an audit item but requires complete review of the IND[*] from the D.C. Regulatory Affairs office. Our final report will include this recommendation.

"The Oxy-Fox Inhalers used in the NDA clinical studies were not validated at the time of release. Though the validations were completed concurrently, it is not acceptable. You took a risk in providing product to the OB-Gyno investigators that may not have been safe or effective for human administration. This is an audit observation.

"The Design Reviews did not include representatives from the Engineering or Quality organizations. Agency requirements specifically require attendance of all functions concerned with the design— obviously including Engineering and Quality—to ensure that the final product satisfies the design inputs for the intended use and the needs of the user. It was not done and resulted in an inhaler design that was not appropriately tested. The argument presented by your Program Manager references the NDA clinicals as the required testing. The NDA is not meant to be used to test product designs—preclinical product development is for this purpose. This is an audit observation.

"The product insert is inadequate. As a cardiologist I know, as well as you should, that an overdose of tricyclics can result in tachycardia. The university studies claimed that for the case of the foxepin drug it would not be the case—but the claim is only for oral administration, not for intramuscular injection or for administration by nasal spray. The Medical Event involved a young woman who exercised quite heavily and had indications of hypoaldosteronism. The product insert does not address any interactions or warnings—this is an audit observation. Our final report will recommend to the Regulatory Affairs office that the entire NDA be reviewed.

[*]IND  = Investigational New Drug; NDA = New Drug Application.

"The Agency review team has a basic disagreement with how the entire Oxy-Fox program was managed. You have a defiant Program Manager who appears to have his own agenda. Was he not promoted to Director following the completion of this launch? During the next week I'll consult with Agency experts to determine if that is an observation or not. In my opinion, both the design and program launch teams made decisions introducing substantial risk to the safe and effective development of the inhaler ... only to reduce the development time and assure the program was completed on schedule."

Dr. Slone looked directly at Dr. Vickory, "And I am shocked that you would even allow this to go on. Dr. Vickory, you have an outstanding reputation in cardiac research and pharmacology interactions ... I personally expect you to do better. And I'm sure your organization discovered the mislabeling in the internal investigation and did nothing to recall the product. In my opinion, this is criminal."

Slone and Sawyer were done. The two stood and left the room. Both Susan and Gail remained, forgetting they needed to escort both men to the gate. Gail had expected a damaging report. Dr. Vickory was somewhat surprised.

\*\*\*

Following the end-of-day wrap-up meeting, Dr. Slone placed a call to Angelo.

"Angelo, I'm taking a break from the investigation and will be in town tomorrow afternoon. Meet me at the club around three."

Angelo answered, "I can meet tomorrow but can you give me an update now?"

Slone just said "No . . ." and hung up.

# 26

# KINNEN MANAGEMENT REVIEW

The meeting is being held in the executive conference room. Though the chairs are designed for comfort there is no one in the room feeling comfortable.

**In Attendance**

M.V.Brooks, Kinnen CEO

M.Cohn, Executive VP Quality and Compliance

R.L. Siegal, Pharma President

Dr. Tom Watson, Pharma VP Research and Development

Dr. Susan Vickory, Pharma VP Medical Affairs

Larry Fletcher, Pharma VP Operations

Janet Weatherbe, Corporate VP Human Relations

Gail Strom PhD, Pharma VP Quality and Compliance

Brooks opened the meeting. "No agenda. Gail, give us a summary of the findings so far."

Gail remained sitting and looked around the room. She made eye contact with Larry Fletcher. Larry met her gaze but only for a moment.

*A History of a cGMP Medical Event Investigation*, First Edition. Michael A. Brown.
© 2013 John Wiley & Sons, Inc. Published 2013 by John Wiley & Sons, Inc.

Gail began, "Sue and I met with Drs. Slone and Sawyer from the Agency earlier today. The Agency team will be off-site for a week. I expect the Agency will use this break to review the findings so far. We are at risk of a warning letter for a number of violations of current good manufacturing practice . . . cGMP . . . in control of nonconforming product, lot control, issues with our testing lab, employee training, and inadequate product development testing. In fact I don't expect the team to return. It was obvious from their questions and focus that they were here solely to investigate the Medical Event received in Baltimore. I also expect that the remainder of the audit will be turned over to the local Agency office. To make this matter even worse, Dr. Slone will recommend that the D.C. Regulatory Affairs office review both our IND* and NDA."

Gail hesitated for a moment, thinking whether to bring up the Agency's Medical Event conclusion. She continued, "As a result of the mixup in the filled lots, the Agency may determine the inhaler as the cause of death and this could result in a criminal investigation.

"As I'm sure you know, the implicated lot will be considered mislabeled product. Mislabeling is a major violation. I've put all finished goods on quality hold subject to reinspection. I expect the Agency will require this lot to be recalled.

"Dr. Slone's final request was for our sales information and the educational series given to physicians . . . and their names."

Gail had kept her boss, Max Cohn, up-to-date through an early morning meeting and a daily memo. Max was responsible for the overall Kinnen Quality System.

Max asked Gail, "I know we covered these items and your team did the best they could under the circumstances. As an organization we need to have a plan to address these observations. Any suggestions?"

"As a matter of fact, Sue Vickory and I met with Tom Watson to determine our next step.

"The cGMP observations are in Larry Fletcher's shop. My organization will review how the quality acceptance is being handled at the plant level and will put controls in place to assure that nonconforming products do not get into finished-goods inventory without appropriate testing and documentation. With Larry's cooperation it can be done in a matter of a few weeks.

"Another point is our Program Managers do not have the appropriate scientific training to manage pharmaceutical programs. I'll meet with Janet Weatherbe and review the employee training procedures for

*IND = Investigational New Drug; NDA = New Drug Application.

both our hourly and professional employees. Janet has assured me that her group can organize a program to document that all our people are qualified for their job responsibilities . . . including our Program Managers. If there are deficiencies, Janet's organization, along with my compliance group, will provide the required training.

"This is in Larry's shop as well. Not to point fingers but the Oxy-Fox market launch Program Manager, Keith Carlisle, did everything he could to really piss off the Agency auditors. I find his actions totally unacceptable and embarrassing."

Larry actually jumped up, startling Gail. "I've told you before to stay out of my area. Mark . . . I find Gail's attitude toward Operations totally unacceptable."

Mark Brooks looked at Larry and said, "Larry, please sit down. We'll meet after the meeting and talk. No one in this room, except maybe Gail, is holding you responsible for the audit results. Gail is sharing her take on the audit. We'll fix these items before the warning letter is even received . . . that is, if we even receive a letter. And Gail, please keep this positive . . . no more accusations."

Gail continued, "As I said, the quality observations that Dr. Slone shared with us can be handled.

"A real concern is the Agency Regulatory Affairs review of both the IND and NDA. Tom, Susan, and I want Dan Garvey from R&D and Jim Gonzales from Medical Affairs to start a complete DMAIC* of the foxepin product . . . using the university studies as a start. We can bring in the university staff as consultants to assist. Dan will lead this group and it will be his first priority. If we present a plan to the Agency to do animal toxicity studies with a foxepin intravenous and intramuscular injection, we may be able to keep the product on the market.

"Mark, maybe you can get an unofficial opinion from your D.C. contacts if the drug can stay on the market while the toxicity studies are being done.

"Dr. Gonzales is rewriting the product insert to include interactions, precautions, and usage restrictions based on a typical tricyclic antidepressant compound. He'll take no more than three days to do it and the internal review and approval can be done in a matter of a few weeks. The new insert will have to go to the Agency for approval, and we may want to voluntarily pull the Oxy-Fox Inhaler off the market until the new insert is complete. Just my opinion . . . it is a short time and it will look good in the Agency's eyes."

---

*DMAIC stands for design phases Define, Measure, Analyze, Improve, and Control (see Chap. 9).

Max asked, "You mentioned an item with not testing the inhaler. What's this about?"

"The inhaler is a copy of the Katlin device and was not adequately tested in the preclinical stage. Dr. Vickory and I agree that it will be an observation in regard to design control since representatives from neither the Engineering nor Quality group were involved in the design or the testing protocols. And our argument that the clinicals satisfy the required testing is not acceptable. No way to get around it."

Mark Brooks asked, "Gail, what impact does this have on our NDA submission?"

"I don't know. We'll have to wait for a decision from the D.C. Regulatory Affairs office. But if we get the foxepin animal toxicity studies completed, the Agency may accept our product validation testing and clinical studies as objective evidence that the design is safe and effective for nasal administration. The toxicity studies are the key. No guarantee though."

*Gail thought since she had the attention of the major Kinnen players she would bring up one of her pet concerns.* "Since I have your attention I want to go over the management bonus plans. These plans are structured solely on financial performance. There are no quality or 'on-market product performance' criteria. These bonus plans are systemic to why we are in this situation now. The organization needs to review the plans with a tie to the Quality Policy."

Gail was finished and said, "That's it in a nutshell. We've work to do."

Brooks stood up to address the group, "Gail's plan appears sound. My only input is we are not pulling the inhaler off the market voluntarily nor will the bonus plans change. You all have to understand that the bonus plans are based on growth and stock performance as our primary concern. If the Agency pulls the Oxy-Fox off the market . . . let them, it's only one product. The rest of her points are well taken . . . we have work to do. Larry . . . we need to talk later and I would like Janet Weatherbe to attend."

*Brooks made a mental note to have Susan DeAngle, Pharma VP Marketing and Sales, in his office first thing in the morning. Giving sales information and the educational series to the Agency could be a problem.*

\*\*\*

Following Brooks' meeting with Susan DeAngle, he met with Janet and Larry. Neither meeting was scheduled. Brooks was annoyed that these meetings would prolong his day in the office; his son had a baseball game that evening he didn't want to miss.

The meeting was informal. Brooks started off, "I know it's early. I usually come in at this time to get myself organized for the day. There's a coffee machine in the break room if you want a cup . . . coffee's not too bad. I want to review Keith Carlisle's participation in the audit."

Larry interrupted, "Mark, Keith is my top performer. He brings to the table an unvarying focus on meeting schedules and that is what we need. I support his actions on the Oxy-Fox product launch."

"Larry, the entire company will be under Agency scrutiny as a result of this audit. Keith's actions in the meetings with the Agency are an embarrassment as Gail pointed out . . . very unprofessional. Keith was belligerent and brought unnecessary attention to our procedures. He must be dealt with.

"Janet, I want you to put together a separation package and I want him out of the company by the end of the month. Keep this strictly between us. When you have the package together just show up in his office and give him two hours to vacate the premises. You can use the excuse that his position has been eliminated in the reorganization. We can use his separation to our benefit with the Agency.

"That's it. Janet you can leave. Larry, stick around."

As Janet left the office, she felt that heads would fall. Blame had to be given and she didn't think Keith Carlisle would be the only one.

"Larry, we need to review your organization. The consultants have moved some of your responsibility to different areas and consolidated the plant and support groups. This audit has made me painfully aware that as an organization we need to put more of an emphasis on the quality of our products. We must get the plant quality procedures in line with Agency cGMP regs. This is your first priority and you will do whatever Gail says has to be done. I want a more visible Quality presence in your operations. Also, the operations Quality function headed by Marcia Hines will report directly to Gail effective immediately. Is it understood?"

Larry answered, "Yes I understand. Mark . . . I've followed your lead in getting our job done regardless of Agency requirements."

"Larry, dealing with the Agency is at my level, not yours . . . I've stonewalled the Agency on minor plant audits but as I just said, if our company is to survive, we must design quality into our products and maintain effective operational control. We're involved in a patient's death and I suspect my Agency contact will avoid us like the plague. I want a complete review of how Pharma runs these program launches. The Agency brought up some points that we have to consider as poor program management."

Brooks was finished and said, "Have a good day."

When Larry left, Brooks left a message for Carole to set up a meeting with the Pharma President, Bob Siegal, first thing in the morning.

\*\*\*

Bob Siegal was running late and didn't get the meeting announcement until the time had passed. He called Carole and told her that he had a doctor's appointment and could she reschedule. Carole told him Mr. Brooks could be available to meet at the end of the day.

Siegal came a little early and sat in the reception area outside Brooks' office. It is one of the nicer areas in the entire complex. The area is roomy with comfortable chairs and cordoned off from the other executive suites. His office—in the operations building—is not nearly as nice. Siegal had a feeling he knew what this was about.

When Brooks arrived, he invited Bob into his office and closed the door.

The Chief Executive's office is furnished at the level appropriate to his responsibility. The walls are the same brick as the exterior of the building, with windows overlooking the entire Kinnen site. He even has his own bathroom. A meeting area to the side is expensively furnished to impress the visitors he entertains. Mark invited him to have a seat. They sat in the meeting area.

"I want to get right to the point. I put you in the position as Pharma President based on your prior financial leadership. Times are different and we need a strong leader who can direct the growth of Pharma from an internal quality perspective. You aren't equipped to do this function. I suggest you submit paperwork to Human Resources announcing retirement effective on your anniversary date. It is three months from now. You can keep it confidential if you like and maintain the office until that date."

"Mark, I've followed your direction in setting Pharma goals. Quality has not been one. I find this to be totally unacceptable and ask you to reconsider."

"Bob, I've made my decision. If you do not retire, I will have you fired based on the results of the Oxy-Fox audit, of which you had no presence. Is that clear?"

Siegal just answered, "Yes."

# PART SIX

# RECKONING

# 27

# BLAME AND RESPONSIBILITY

## 27.1 THE INVESTIGATION IS A PUBLIC RECORD

The next afternoon Bill Slone met with Angelo Walden. Bill had a table reserved in a quiet part of the club dining room and had arranged that no one would be seated within earshot.

Angelo was late; the time gave Bill the opportunity to get his thoughts in order. He would go over the audit, with highlights of the observations his team had determined. But what was Angelo going to do with this?

Angelo walked through the door, acknowledging the dinning room staff and members he knew. Angelo sat next to Bill, on his right side.

"Sorry I'm late . . . had trouble getting away in the middle of the day."

"No problem . . . gave me time to think this out. Our audit is complete . . . we have information on the drug we suspect contributed to Francesca's death. Before I go over the data with you I want to know what you're going to do with the information."

"My family needs closure. We need to understand how and why Francesca died. No more than that."

Dr. Slone continued, "Makes no difference anyway. A warning letter will be sent to Kinnen and it is public record as well as the testimony

*A History of a cGMP Medical Event Investigation*, First Edition. Michael A. Brown.
© 2013 John Wiley & Sons, Inc. Published 2013 by John Wiley & Sons, Inc.

and minutes transcribed during the audit. Whether you hear it from me
or read in on the Agency's website doesn't really make a difference.

"I'll just give you highlights. Kinnen is in violation of a number of
quality regulations affecting the dosage of the inhaler that could have
contributed to your niece's death, and the program management team
made a number of shortcuts in the design of the product itself, potentially making the inhaler unsafe. I'm confident that the Agency will
require the drug to be taken off the market until these observations
are corrected.

"In addition, the audit led the team into a probable physician sales
kickback scheme and insurance fraud. This practice is unethical but not
necessarily illegal. We'll turn our findings over to the Justice Department for investigation. The Agency has no role in this area."

Angelo listened and weighed his question carefully, "Can you give
me names of anyone involved?"

"Angelo, the people from Kinnen will be in the public record
as well."

Dr. Slone hesitated.

Angelo asked, "Is that it?"

"No ... There are a few high-level managers at Kinnen and, unfortunately, in the Agency's D.C. office who I personally hold culpable for
the entire episode."

Angelo was getting what he wanted, "And their names?"

## 27.2  KINNEN WRAP-UP

Gail Strom was correct, the Baltimore audit team did not return. The
audit was turned over to the local Agency office and would continue.
A registered letter was received to that effect. The letter listed the
cGMP* observations that Dr. Slone covered on the last day of his visit.
Kinnen was prepared. The activities were in place and well on the way.
The warning letter gave Kinnen fifteen calendar days to respond with
a plan to correct the quality observations. A second registered letter
was received from the Agency's D.C. Regulatory Affairs office notify-
ing Kinnen that both the IND† and NDA were being reviewed by the
Agency and, until the review was complete, the Oxy-Fox Inhaler must
be removed from the market and all product recalled. This result was
also as anticipated.

*cGMP = current good manufacturing practice.
†IND = Investigational New Drug; NDA = New Drug Application.

Brooks tried to do damage control through Dr. Louis Koppel, Agency Deputy Commissioner, but as Brooks surmised, Koppel treated him as if he did have the plague. Dr. Koppel was more concerned with an Agency internal investigation instituted by his Baltimore office than the fate of Kinnen.

At the end of the month, Janet Weatherbe showed up in Keith Carlisle's office accompanied by two security guards. The conversation was short and to the point.

"Janet, when you show up with Security I expect someone is getting fired."

*Keith considered himself a valuable asset to the Pharma Division . . . though the Oxy-Fox Inhaler was somewhat of a fiasco. But it wasn't his doing . . . if the teams hadn't disagreed with him so many times, the audit wouldn't have ended the way it did. He had met his personal goals . . . MBA and Director Program Management. The new bonus level had paid for his new BMW 530i. He thought for a moment . . . "who could HR be after."* His ego would never have recognized that HR was after him.

Janet sat down without invitation; the security guards remained standing. "Keith, I'm afraid I have to inform you that, as a result of the recent reorganization, the position of Director Program Management has been eliminated. The company put together a very attractive separation package for you and we'll assist in your efforts in finding a new position."

Janet paused—she could not believe the look on Keith's face.

"I don't understand. The Director position was a result of the reorganization. I've only been in the position for four months."

"The consultants have done a reassessment and it is the way it is. I'm giving you the package for your review. You've the rest of the day to seek legal counsel if you choose. If you have questions, you must be in my office by 8 AM tomorrow morning . . . you may bring an attorney if you like, and you have fifteen days to sign the separation papers or the package will be rescinded. I strongly suggest you read the information and join me tomorrow morning at 8 AM."

"Janet, I don't believe this. Is Larry aware?"

"Yes. Larry is aware. He would have joined us this morning but he has more pressing affairs."

"Okay. Guess I have no choice. I'll meet you tomorrow morning."

"Keith, one more thing . . . you have two hours to remove any personal items you have and vacate the Kinnen campus. The security guards will remain with you and collect your identification badge as you leave. You may have noticed that your computer has been made inoperable."

\*\*\*

Keith was in total shock. How could Larry let this happen? After all, he was Larry's "go-to guy"!

Larry was trying to save his own job—at least until he found another.

Keith read the separation package carefully. Janet was correct; the financial arrangements would carry him for more than a year. He would find something else even better. No problem meeting her tomorrow and he would not need an attorney. He couldn't pass up the twelve months' salary offered.

# 28

## CLOSURE

Keith Carlisle arose early the next morning. He needed only to wear casual clothes to the meeting with Janet. Through contacts with other healthcare companies and with his MBA and Director Program Management title, he would find an even better position within a few months. For the time being he'd take it easy and enjoy some of the benefits that his past bonus and separation payment afforded.

Keith ate a leisurely breakfast of eggs and bacon with Greek toast—he liked the texture and thickness of this toast—leaving the dishes in the sink. He didn't live far from Kinnen, about a fifteen-minute ride in his new Beemer.

Keith had dinner with his girlfriend of the month the night before to celebrate his financial windfall and had come home late, leaving the car in the driveway. He lived in a complex of town homes, each with an attached garage separated by a private walkway. Ever since he bought the new Beemer he left the car in the driveway anyway. The neighborhood was safe and he liked showing off his success to the neighbors.

Keith left the house through the front door. His thoughts were on the money; a sailboat would be nice, he always enjoyed sailing but never had the time to do more than crew someone else's boat.

*A History of a cGMP Medical Event Investigation*, First Edition. Michael A. Brown.
© 2013 John Wiley & Sons, Inc. Published 2013 by John Wiley & Sons, Inc.

He was oblivious to the dark figure standing at the side of the garage, hidden in the walkway—watching, waiting, and carefully observing any activity in the surrounding homes. As there was none the dark figure made a move.

Keith walked to the driver's side of the car and pushed the remote to open the door and start the engine. The stranger stepped out and confronted Keith before he could get into the car.

Keith looked up and saw the delivery end of a shotgun pointed directly at his face.

Keith was frozen with fear. The assassin said four words, more than should have been said, "This is for Francesca" and paused slightly before pulling the trigger. In the split second of recognition Keith thought, "...*who is Francesca?* ..." Keith hit the ground hard—the pellets shattered the base of his brain—Keith would never get up again.

\*\*\*

The assassin turned and slowly left through the walkway and small back yard, exiting on the side street. A car was waiting. The passenger-side door opened.

There was no reason for the two occupants to speed away from the scene, drawing attention; the car was untraceable. Anthony Walden drove the car and asked his companion how it felt.

Taking off the ski mask and unbuttoning the tight collar of the loose-fitting overshirt, though leaving on the two sets of surgical gloves, his companion signed deeply and said, "...Closure."

Anthony looked at his companion and said, "Mrs. Bucco, he was only the first."

# BIBLIOGRAPHY

AAMI. *The Quality System Compendium: GMP Requirements & Industry Practice*. Arlington, VA: Association for the Advancement of Medical Instrumentation, 1997.

Basem El-Haik, Khalid S. Mekki. *Medical Device Design for Six Sigma: A Road Map for Safety and Effectiveness*. Hoboken, NJ: John Wiley, 2008.

*CSSBB PRIMER*. Quality Council of Indiana, 2001.

FDA. Drug Applications and Current Good Manufacturing Practice (cGMP) Regulations: Parts 210 and 211. Washington, DC: U.S. Food and Drug Administration. Available at www.fda.gov (www.fda.gov/Drugs/DevelopmentApprovalProcess/Manufacturing/ucm090016.htm).

FDA. 510(k) Submission for Medical Devices. Washington, DC: U.S. Food and Drug Administration. Available at www.fda510k.com/approval-process.

FDA. Guidance, Compliance, & Regulatory Information. Washington, DC: U.S. Food and Drug Administration. Available at www.fda.gov (www.fda.gov/cder/regulatory/default.htm).

FDA. Investigational New Drug (IND) Application. Washington, DC: U.S. Food and Drug Administration. Available at www.fda.gov (www.fda.gov/cder/regulatory/applications/ind_pa1.htm).

FDA. New Drug Application (NDA). Washington, DC: U.S. Food and Drug Administration. Available at www.fda.gov (www.fda.gov/cder/regulatory/applications/nda.htm).

---

*A History of a cGMP Medical Event Investigation*, First Edition. Michael A. Brown.
© 2013 John Wiley & Sons, Inc. Published 2013 by John Wiley & Sons, Inc.

FDA. Title 21 Code of Federal Regulations Part 58. Good Laboratory Practice Regulations. Washington, DC: U.S. Food and Drug Administration. Available at www.fda.gov (www.fda.gov/ohrms/dockets/98fr/980335s1.PDF). [21CFR58]

ISPE. *GAMP Guide for Validation of Automatic Systems*. Tampa, FL: International Society of Pharmaceutical Engineers (ISPE), 2002.

# INDEX

Page numbers in *italics* refer to figures.

*A History of a cGMP Medical Event Investigation*, First Edition. Michael A. Brown.
© 2013 John Wiley & Sons, Inc. Published 2013 by John Wiley & Sons, Inc.

meeting format, 50
product specifications, 111–112
project charter, 35–36, *37–39*
quality functional deployment, 99
Technical Feasibility Document,
    41
tollgates, 41, 77, 98, 103, 118
voice of the customer, 40, 97, 98,
    111, 112
Slone, William (FDA investigator)
conversation with Walden, 150,
    152–154, 209–210
due-diligence report, Katlin
    studies and design history,
    192–195
FDA arrival, on Kinnen site,
    183–186
FDA audit observations, 197–199
role of, 179
Social behavior. *See* Behavior;
    Postpartum depression
Sodium ion, 147
Sophia (childcare provider)
death of Francesca, 141, 142
Medical Event morning, 6, 7
medical information provided for
    autopsy report, 147, 148
role of, 5
Stall, Dave (sourced materials
    manager)
cGMP process validation meeting,
    69
due-diligence team, 24
failure mode effects analysis
    meeting, 81, 89
FDA investigation, 193
fishbone diagram meeting, 105
launch team meetings
    first meeting, 27, 28, 42
    second meeting, 49
    third meeting, 55, 60–61
management structure, 24
product development and
    manufacturing meeting, 66
role of, 27

Start-up issues, 129–135
Stevens, Barry (engineer), 123
Strom, Gail (quality VP)
audit letter meeting, 180–181
audit results, 210–212
DMAIC investigation process,
    163–170
executive management review,
    174, 175
FDA investigation
    arrival of, on Kinnen site,
        183–185
    audit observations, 197, 199
    due-diligence report, Katlin
        studies and design history, 193
FDA reporting, 170
final management review, 135,
    136
internal quality review, 171–173
investigation team meeting,
    161–162
management review meeting, 201,
    202–204
Medical Event report, 158–160
Summers, Elisabeth (obstetrician), 4,
    144, 145
Symms, Ronald (FDA cGMP
    expert)
NDA clinical studies review,
    189–190
qualification and quality
    acceptance records, 187–189
System FMEAs, 86

Tachycardia
in humans, 148, 160, 198
transgenic mouse study, 18
tricyclic compounds, 16
Taylor, Gordon (product
    development)
cGMP process validation meeting,
    69, 70
design team meeting, 115, 116, 117,
    118
discrepancies noticed by, 24

Printed in the United States of America
ED-12-15-12